SpringerBriefs in Political Science

More information about this series at http://www.springer.com/series/8871

Trevor Brown

The Digital Galactic Complex

Springer

Trevor Brown
Lake Geneva
WI
USA

ISSN 2191-5466 ISSN 2191-5474 (electronic)
SpringerBriefs in Political Science
ISBN 978-3-319-22385-8 ISBN 978-3-319-22386-5 (eBook)
DOI 10.1007/978-3-319-22386-5

Library of Congress Control Number: 2015947935

Springer Cham Heidelberg New York Dordrecht London

Printed on acid-free paper

Springer International Publishing AG Switzerland is part of Springer Science+Business Media
(www.springer.com)

Contents

1 Introduction—A Space Joint Stock Company 1

2 The First Division—Security Wing 9
SSP 9
A Debris Service 13
Intelligence 18
Cyber 19

3 The Second Division—Space Colonization 21
In Space Manufacturing 21
The Moon 26
Comets and Asteroids 29
Mars and Beyond 33

4 The Third Division—Spillover 37
The Effect of Technology Development on the Broader Economy 38
R&D 39
Societal Effects of R&D 42
Space and Environmental Symbiosis 42
Revolutionary Technologies Flowing from R&D 43
Speeding up Technological Advance 46
The R&D Process 46
The Interrelationship Between the Space and Global Economies 48
The Company's Social Altruism 50

5 The Fourth Division—Finance 53
The Capital Markets 53
Factors in Valuing the Market 56
The Company's Strategy on the Worldwide Equity Markets 59
The Fourth Division's Central Banking Function 61

6 Grand Strategy 69
Europe 69
Asia 71
The Middle East 74
The Global South 75

7 Conclusion 79

References 81

Chapter 1
Introduction—A Space Joint Stock Company

Space civilization could mean the onset of a new epoch. Indeed, the duration of the days and years for the planets and their moons are different than for those of the Earth which will necessitate a new system by which humanity keeps time when these celestial bodies are colonized. It is a very auspicious period that human civilization is entering. Thus far the national space effort of the United States has gotten to this point by incremental steps. With the exception of the Apollo program, a grandiose achievement no doubt, the American national space effort has proceeded with small adjustments to the policies of its successive administrations over the decades.

In this regard, on June 28, 2010, the Obama administration released a new policy to guide the national space efforts of the United States (Obama Policy 2010). Observers noticed a "change in tone" from the previous administration's space policy (Smith 2011, p. 20). The new Obama policy emphasized increased cooperative efforts in space between the United States and the international community in contrast to the unilateral positions of the previous administration (Smith 2011). However, while the tone of the two policies was quite different, which was reflected in the manner that each of the administrations was conducting their affairs, the core principles of each of the documents, although phrased differently, were largely similar (Kueter 2010). As such, there has been a large degree of continuity in the main principles guiding the American national space effort over the various administrations. Nowhere this was more evident than the successive American administration's policies to leverage the dynamics of the private sector in their respective space policies. Since the Reagan administration of successive American administrations has increasingly sought to draw the private sector into

T. Brown, *The Digital Galactic Complex*, SpringerBriefs in Political Science 30,
DOI 10.1007/978-3-319-22386-5_1

their space policy formulations, this was the case when it came to the privatization of the American civil space effort. In this regard, the Bush policy stated

> The United States is committed to encouraging and facilitating a growing and entrepreneurial US commercial space sector. Toward that end, the United States Government will use US commercial space capabilities to the maximum practical extent, consistent with national security (Bush Policy 2006).

The Obama administration furthered this policy when on February 1, 2010 it announced "as part of its 2011 budget request" that it would "rely on the private sector" to transport crew and cargo to low Earth orbit (LEO) rather than undertaking to develop a NASA vehicle to perform these transportation services (Smith 2011, p. 22). The NASA vehicle was to be a variant of Constellation for LEO travel. Under this new policy, the federal government would provide "$6 billion over 5 years" to private operators to develop spacecraft for LEO travel (Smith 2011, p. 22). Once the private sector spacecraft was ready, the government would then purchase space on them for astronauts and supplies. It is important to note that the government funds were not intended to pay for the full cost of the private spacecraft. It was only meant to boost their research and development (R&D) efforts. The private operators would then make profits by ferrying crew and cargo to the International Space Station (ISS) as well as to additional infrastructure in LEO that was being developed elsewhere by the private sector. It was thought that this arrangement would allow NASA to devote more resources to human spaceflight and scientific efforts in deeper space, such as a mission to an asteroid by 2025 (Smith 2011).

Concerning a mission to an asteroid, while Obama's asteroid redirect mission did represent a radical departure from Bush's vision to return to the Moon, it remains to be seen if the asteroid mission will be carried through by future administrations. Meanwhile, the policies that actually do get implemented, such as commercial crew and cargo, definitely do represent incremental adjustments to the policies of previous administrations.

Nevertheless, the key challenge facing the Obama administration in regard to LEO was how to "successfully transition from the Shuttle to the Constellation program" when scarce resources were available (Sadeh 2009, p. 110). Ultimately, the Obama administration dropped plans for Constellation and attempted to energize the private sector with an infusion of government funds. It was thought that this would allow NASA to maintain "the current and planned programs and projects in space and Earth sciences and utilizing the ISS once it is fully assembled" (Sadeh 2009, p. 110).

The important point to observe is that at the time of the Obama administration's policy formulations, the American private sector was beginning to demonstrate a critical competence for LEO space travel. As the private sector was about to begin transportation services for LEO, it would have been foolish from a policy standpoint to duplicate the private sector's efforts by continuing to develop an expensive spacecraft when the private sector operators could perform the same service far more economically. As such, there are a variety of other areas in the American national space effort that has the potential for the same duplication. Therefore, perhaps it would be a natural evolution of the present and future American administrations' space policies to extend the industry–government partnership for LEO space travel to these efforts and encourage the establishment of an organization that would be comprised entrepreneurs from Silicon Valley and NASA personnel. This new enterprise would be private but would receive a measure of financing from the federal government in the same manner as the policy for commercial crew and cargo. The organization would take over the space effort that was previously led by NASA on the behalf of the United States government, and NASA would gradually evolve into a regulatory agency. Indeed, while up to this point, the American national space effort has evolved by incremental steps such a policy would poise the national space effort to leap by revolutionary bounds.

Regarding duplication between the space efforts of government and industry, NASA and the United States military have been interested in space solar power (SSP) for some time (Jennings 2014; NASA 2014) while a number of private SSP companies have been created. SSP is the concept of collecting solar energy on orbit and transmitting it to rectennas on Earth to be delivered into the electrical grid. Since 2009, engineers at the Naval Research Laboratory (NRL) have been working on a solar power satellite (SPS), the design of which they specified as the "sandwich module" (Jennings 2014). The concept for the SPS has been around for decades but has thus far proven elusive to its advocates because there has previously not existed the technical and financial means to create structures with extremely large surface areas in space (Brown 2009). The cost to launch the necessary materials and the cost of the in-space assembly and supporting infrastructure have proven prohibitive, while the techniques necessary to perform the construction have proven technically out of reach as well (Brown 2009). Unfortunately, this was the case with the NRL's most recent design. The design called for the assembly of 80,000 pieces that would be 10 ft on a side each for a structure that would have been about nine football fields wide (Jennings 2014). Obviously, such an effort would have required thousands of launches for what would have been at most 5 GW (Jennings 2014). Indeed, it is still unclear if the design would have required an even larger structure of mirrors to produce even that much energy. Without such mirrors, the design may have only produced hundreds of megawatts. Regardless, with so much assembly required (80,000 pieces), it would have been necessary for the NRL to invent a significant amount of extremely sophisticated robotic technologies to perform all of the assembly, which would have been a difficult proposition at best while increasing the costs substantially (McDuffee 2014). There were similar issues for NASA's most recent design called the SPS Alpha. The SPS Alpha would require the assembly of tens of thousands of small elements in space with robots for a structure that would generate just a few gigawatts at most (NASA 2014). The SPS Alpha is just the latest design in a long history of NASA's stunted attempts to study and develop solar power satellites. An issue for these efforts was the ability to construct a large enough surface area in orbit to generate meaningful amounts of energy for markets on the Earth. In this regard, NASA and the NRL, like all proponents of SSP, are at an impasse.

However, there are revolutionary materials that have already been developed that would allow NASA and industry to overcome this impasse. L'Garde, in combination with NASA's Jet Propulsion Laboratory and Northrop Grumman, has already invented the inflatable photovoltaic cell (Lichodziejewski et al. 2002). While this material would require additional refinements, the concept of inflating the photovoltaic surface area, rather than painstakingly trying to assemble it on orbit, will allow former NASA personnel and Silicon Valley to loft massive surface areas into space (Brown 2009). The surface area of the asset would be photovoltaic and would collect solar energy on orbit. But by inflating the satellite, former NASA personnel and Silicon Valley could overcome the mass and volume constraints of existing and future rockets. Indeed, engineers and contractors at NASA's facilities in Huntsville, AL have created inflatable solar arrays which would allow for the generation of at least 1 KW of energy in a stowage volume that would otherwise allow for only tens of watts with conventional arrays like NASA's and the NRL's most recent SSP efforts (NASA 2014). These technicians have already demonstrated that the inflatable concept will allow former NASA personnel and Silicon Valley to loft massive surface areas into space with absolutely no assembly whatsoever. What is more, to produce a demonstration satellite for this concept would require very little time and effort for former NASA personnel and Silicon Valley. L'Garde could produce a 1 km in diameter inflatable photovoltaic structure almost immediately. This inflatable would also be as big as NASA's and the NRL's previous designs which would require the assembly of tens of thousands of elements on orbit. Indeed, L'Garde has indicated that this asset could generate at least 200 MW in space and could be orbited with a launch cost of $1 billion. Therefore, it most definitely is conceivable that if a small aerospace company such as L'Garde could produce such an asset on orbit, then the nation's most sophisticated labs, in combination with the nation's most sophisticated aerospace companies, could also orbit in one launch with an existing rocket a 1 km inflatable that would generate 200 MW in space, and that which, critically, would involve no assembly. Without doubt if L'Garde could easily and quickly develop and constitute such an asset on orbit, then the nation's most sophisticated labs, in combination with the nation's most sophisticated aerospace companies, could certainly make the asset lighter, stronger, and much bigger with the billions of dollars that they would surely have for R&D from the federal government. One would think that former NASA personnel and Silicon Valley would not have too much trouble in this regard.

As far as designs go, planar models would solve the problem of the Beta angle, but the simplicity of spherical spacecraft could prove more desirable as they would not have to adjust to remain in view of the Sun. As these satellites would have to be many kilometers long, spherical designs could reduce the complexity of the spacecraft considerably (Brown 2012). Therefore, the spherical design would mean that there would be very little station keeping.

Concerning the threat of micrometeorites, as the medium of space is a vacuum, the structure would not "pop" if a small hole gets punctured in it. The electrical charge that was used to inflate the structure would help it keep its shape and maintain integrity on orbit even if it were getting struck by micrometeorites. As such, the structure would continue to generate significant energy even with minor damage. Thus, the inflatable would likely be even more survivable than NASA's and the NRL's previous designs.

Concerning the issue of the degradation of solar cells, L'Garde's photovoltaic technology is 30 % efficient at the present time (L'Garde 2014). It is conceivable that with NASA's and Silicon Valley's extremely capable personnel, further developments in this technology could lead to 40 % efficient material. As such, former NASA personnel and Silicon Valley could conceivably produce an asset that would average at least 20 % efficient solar cells over the course of a 10-year period making such technology extremely viable economically.

Concerning energy transmission from orbit to the surface of the Earth,

> power can be transferred (beamed) directly to where it is needed. The technologies to do this are not magic or unfamiliar—they are the same elements used every day to emplace, power, and communicate with every existing satellite. In a mature system-of-systems, multiple solar-power satellites would reside in geostationary orbit, each collecting vast amounts of power and transmitting it through active electronic beam steering, like routers in a vast orbiting power internet. While appearing to hover above a particular location, each SPS could service multiple markets, providing power on demand to urban centers or remote locations. For example, a single satellite south of Baja California could service markets across most of North and South America; a satellite over the Indian Ocean could service markets as far apart as Africa and Indonesia, and from Diego Garcia to as far north as Russia. Power in this system-of-systems would be transmitted using a technique called retrodirective phased array, where an encrypted pilot signal from the ground handshakes with the satellite's active electronic beam-steering system to link transmitter and receiver. The beam itself would be in the ISM band (typically 2.45 or 5.8 GHz), so that it passes nearly full strength through the atmosphere, clouds, and rain. Because of low atmospheric losses (<2 %), extremely efficient reconversion (>80 %), and most of all, constant illumination, the beam can be safely kept at an amazingly low intensity (only one-sixth the intensity of sunlight) and yet be significantly more energy productive than a comparably sized terrestrial solar plant. The location and diameter of the beam are predictable and well confined" (Garretson 2012, pp. 99–100).

At this point in the space game, the idea to inflate structures on orbit should be self-evident. Even SSP should become a serious proposition if the assets are about to be easily inflated just like Bigelow Aerospace's habitats, and all of the other antennas in space that a multitude of space actors are proliferating. Indeed, inflating a photovoltaic asset on orbit that would not have to support human life would likely be technically easier than it was to produce the man-rated habitats that Bigelow Aerospace is about to orbit in the next few years. What is more, one private sector SSP company, Powersat, has already proposed inflatable SSP designs. While Powersat ran into problems raising money for their design because questions about station keeping and beam pointing could not easily be answered, NASA and the United State military already have the personnel to easily overcome such issues. As previously discussed, station keeping will not be much of an issue because of the assets' spherical design, and engineers at the NRL can already point the beam (Jennings 2014). As such, former NASA personnel and Silicon Valley could easily draw on these personnel and would be poised to move out with

SSP in the immediate future. With all of the developments that have already gone into the production of inflatable photovoltaic cells and energy transmission, former NASA personnel and Silicon Valley could easily be on orbit with inflatable solar power satellites in a matter of years and not decades as would be the case with all of the conventional designs being developed elsewhere around the world.

For both the recent Bush and Obama administrations, energy has been, perhaps, their most salient issue. Indeed, the Obama administration has been touting ground-based solar initiatives throughout their time in office. And while this has been the case, the American Congress has appropriated substantial sums of money for interests in the American solar industry in recent years. Meanwhile, a number of structures, from habitats to massive antennas, will be inflated in space in the immediate future. As such, it would now be a logical evolution of the American government's policies to extend the policy for LEO space travel to SSP and constitute an industry–government partnership that would inflate a solar initiative in space. It would simply be a matter of making the inflatable solar arrays that NASA is already developing as big as possible.

The backdrop for these developments is a nation, the United States, in relative decline compared to the nations of Asia. Various states around the globe have giddily noted this dynamic. The 2008 recession took a devastating toll on the national economy and the finances of the United States government. Following this event, the fiscal realities facing the United States government became bleaker and bleaker. The debt surged by trillions. This prompted sequestration—across the board government budget cuts—which have had a tremendous impact on the Pentagon's and NASA's budgets. Indeed, over the course of the present decade, the Pentagon is planning to cut $1 trillion from its budget proposals. NASA's budget has barely grown since 2012. NASA now celebrates when its budget merely does not get cut, even if it barely grows.

And yet while the fiscal situation has been this abysmal for the United States government, it is only going to go from bad to worse with a massive demographic shift on the horizon. At the end of the decade the baby boomers will be moving into retirement, which will put a devastating strain on the federal budget. The needs of entitlement programs such as Social Security and Medicare will explode. Members of Congress have been proposing drastic reforms to these entitlements already. Astute observers can believe that before the benefits of these entitlements are cut, the Pentagon's and NASA's budgets will be devastated. When this fiscal reality sets in, the Pentagon and NASA could become shells of their former selves.

However, as Twain once remarked, rumors of our demise are greatly exaggerated. In fact, with inflatable photovoltaic structures in play, the prospect of conquest and riches is holding forth in the heavens. As has been discussed, if enterprising members of Silicon Valley and NASA took the initiative to create an industry–government partnership in the form of a Space Joint Stock Company, with the inflatable SPS as its centerpiece, the United States could rise to a zenith of wealth and power that would exceed anytime in its previous history (Brown 2012). The Space Joint Stock Company would be similar to the famed British East India Company of centuries ago in which it would be constituted to explore for natural resources and would develop colonies in support of this mission. The difference being in the medium—whereas the British East India Company's operations were in the Orient, the Space Joint Stock Company's operations would be across the solar system. The Space Joint Stock Company would also be similar to the East India Company in which it would contain a security wing to provide for the protection of its commerce and possessions.

In addition to entrepreneurs from Silicon Valley, perhaps the Space Joint Stock Company would be backed by a consortium of companies from the American aerospace industry that would hold shares in the company. Boeing, Lockheed Martin, Northrop Grumman, Raytheon, General Dynamics, General Electric, etc. all may find considerable opportunity in such a venture. Indeed, these companies could potentially pool a share of their industrial resources to support the company's operations. In this way, the Joint Stock Company could generate a significant amount of organic growth.

Nevertheless, it might still be necessary for the Joint Stock Company to also receive financial assistance from the United States government in a similar fashion as the private companies have been contracted by NASA to deliver crew and cargo to the International Space Station (ISS). Maybe the Joint Stock Company could even receive hundreds of billions of dollars in aid from the United States government given the scope and importance of the projects it would have ongoing. Perhaps the Silicon Valley entrepreneurs and NASA personnel who would found this venture would even enlarge the consortium to include companies from other American industries such as the energy and IT sectors. Such an initiative could leverage the tremendous synergies to be had from the technical and personnel resources of companies in these very important industries. Quite obviously, companies such as Google and IBM could contribute much of the information technology that the company would utilize, while oil companies such as Exxon Mobil, Chevron, and Conoco Philips could adapt their platforms to extract resources in the cosmos.

The company would be organized along the lines of four divisions. The first division would contain the inflatable solar power satellites and would comprise the security wing of the company. The second division would be constituted to explore the cosmos in search of natural resources and would be tasked with colonizing the inner solar system in support of these missions. The third division would manage the spillover of all the synergistic technologies for space that could be brought to terrestrial markets. And finally, the fourth division would be tasked with managing the immense financial resources of the company by engaging in investment operations, reserve banking, and consumer finance.

The company would have considerable capabilities with which to engage the world. The company's business activities would span the planet, in addition to that of the solar system, but the company could also have a psychological operations campaign (psy-ops) with its assets on orbit to create the international dynamic that would be most conducive to its business. And if the international situation became hostile, the company would be able to protect its interests with its security wing. In any event, the Space Joint Stock Company could usher in the greatest period in the history of human civilization.

Chapter 2
The First Division—Security Wing

SSP

The first feature of the first division of course is the inflatable solar power satellite. Once the 1 km in diameter demo satellite was successful the company could proceed with larger assets. In this regard, in just one launch with a Delta IV the first division could loft a satellite 5 km in diameter into a geosynchronous Earth orbit (GEO). No assembly would be required for a satellite that would then generate over 5 GW on orbit. This energy could then be transmitted to the Earth via microwaves in the aforementioned way, supplying as much or more energy than most nuclear power plants (Brown 2012). With just a few dozen launches with a rocket that is already in existence, the first division could immediately begin to alleviate much of the energy needs of a region of the United States in a very environmentally friendly fashion (Brown 2011).

However, to supply the energy needs of the whole of the country, a much larger satellite would be necessary. In this regard, a satellite 300 km in diameter could generate more than 18 TW on orbit. As this satellite would require some assembly perhaps it would be best to locate it closer to the Earth in a medium Earth orbit (MEO). If there were three such spheres rotating in MEO, then there would be enough energy to not only meet the needs of the whole of the United States, but also even those of the entire world (Brown 2010). Such satellites would appear majestic in the sky and would constitute nothing less than "Solar Moons" (Brown 2012).

The concept of inflating an artificial moonlet in Earth's orbit is definitely viable. In fact, during the Vietnam war NASA proposed to the Department of Defense "that it place an artificial moonlet in synchronous orbit over areas such as Vietnam to aid military operations at night" (Schauer 1976, p. 67). The idea was that the artificial moonlet would reflect light onto the theater of operations. While such an asset was rightfully recognized to be of dubious value by the Pentagon during

T. Brown, *The Digital Galactic Complex*, SpringerBriefs in Political Science 30,
DOI 10.1007/978-3-319-22386-5_2

Vietnam, today assets that could supply the energy needs of the whole of the globe with the cleanest energy available would be the most critical assets in the nation's inventory. In any case, if NASA could have created an artificial moonlet decades ago, surely Silicon Valley, NASA, and the American scientific community could create one today. Indeed, the establishment of three 300 km in diameter artificial moons inflated on orbit would involve the assembly of only a few hundred very large photovoltaic pieces for all three structures together. This would obviously be a dramatic reduction from the tens of thousands of pieces required for the assembly of just one of the solar power satellites that NASA and the NRL have been attempting. Clearly, even massive inflatable structures such as the Solar Moons would be a far easier proposition than NASA's and the United States military's previous designs. What is more, while NASA's and the NRL's previous designs would only be generating a few gigawatts at most the Solar Moons would be generating over 50 TW on orbit collectively—enough to supply the energy needs of the whole of the globe.

But furthermore, a constellation of three massive spheres would likely be easier to manage than hundreds, or thousands, of relatively smaller ones. However, for a variety of technical and security reasons it may be more prudent to have the energy supply distributed. Thus, hundreds or thousands of 5 km satellites ringing the Earth could be the best strategic posture for these assets in space. But then again, even if there were this many, they would not be able to generate as much energy as three 300 km spheres. Perhaps then, the most prudent strategic posture would be to place three Solar Moons in MEO and then surround the globe with a few hundred 5 km spheres in GEO.

Whatever constellation of the first division would decide on the financial opportunities would be limitless. The operations costs of three 300 km spheres would likely be only tens of billions of dollars on an annual basis. Unless a 5 km sphere were serving as a base of operations for the company in space, the operations cost of it would probably be zero as it would most likely be replaced if there were a significant problem. Thus, the operations costs of the whole of the constellation would still be only tens of billions of dollars.

In any event, the cost to develop these systems would be immense but totally manageable. At present, the production costs for ground-based photovoltaics are around $1 per watt (24/7 Wall Street 2011). But according to the United States Department of Energy in 2013, worldwide output for the entire solar industry was just 45 MW. So with just one 5 km in diameter inflatable solar power satellite the worldwide production of photovoltaics would surge by a factor of over 400. If the Joint Stock Company were to orbit hundreds of 5 km assets, the worldwide production of photovoltaics would surge by a factor of tens of thousands or more. Indeed, if the Joint Stock Company were to orbit three 300 km in diameter artificial moons the worldwide production of photovoltaics would surge by a factor of millions. Over 840,000 square kilometers of extremely thin and lightweight photovoltaics would have to be produced for three 300 km in diameter artificial moons. If the Joint Stock Company were to achieve such scale, the production costs of photovoltaics would drop dramatically. This could mean that the Joint Stock

Company could possibly orbit 5 km in diameter assets for hundreds of millions of dollars each and three 300 km in diameter artificial moons for just hundreds of billions of dollars. Quite possibly the total cost for the whole of the energy infrastructure could come in at under one trillion dollars.

As far as financing the development of these systems is concerned, if 1 GW of energy could supply the needs of 750,000 homes (Garretson 2012), then the 5 km in diameter assets could be generating around a billion dollars in profits each. In fact, even with the most conservative performance estimates for the conversion/path loss/reconversion/distribution factor, the 5 km in diameter assets could still generate hundreds of millions of dollars in profits each. So after an initial round of funding from the United States government amounting to a few billion dollars to start the operation and develop the first few 5 km in diameter satellites, the profits that these would then generate could finance the development of more and more assets. In this way, the Joint Stock Company could grow the number of assets on orbit with mostly organic processes. Therefore, after orbiting around one hundred 5 km in diameter assets, the company would then have the resources to move out with the development of the 300 km in diameter artificial moons. Or if the political climate in the United States recognized the urgency with which the artificial moons should be developed, perhaps the United States government would undertake to finance their development in all of their entirety by appropriating a few hundred billion dollars for the effort. Such an outlay could easily be managed in installments over a period of a few years and would speed the development of these systems dramatically. Indeed, when considering the fact that the United States government expended over a trillion dollars for the Iraq war, such financial arrangements would look extremely prudent in comparison.

If the United States government did undertake to finance the development of the demo satellite, the first few 5 km in diameter assets, and even the artificial moons, it would mean that the Silicon Valley entrepreneurs and the elements from American industry that created this organization would have incurred virtually no financial risk in the process. However, while the United States government could provide financial support to the organization, Silicon Valley and American industry would certainly not need it. They already have the resources to create the energy constellation with wholly organic means.

Regardless, the global energy and transportation marketplace is roughly $7 trillion (Garretson 2012). This could mean over one trillion dollars in profits in the first year of the Solar Moons' operation alone. Indeed, as the cleanest energy there is, solar is politically attractive. In addition, the cost of the energy would undercut all global energy competition as well. But then energy consumption would quickly leap with the ensuing economic expansion (Brown 2012).

As such, this energy concern could conceivably achieve 50 percent global market share, which would put revenues at around $3.5 trillion. Some estimates have this concern achieving a near monopoly at 75 % global market share, which would put revenues at in excess of $5 trillion. As the entire slate of development costs would be less than a trillion, and as costs to operate the constellation would be in the tens of billions, then these projected cash flows could give this concern a valuation of $50–$100 trillion. If this concern was based in the United States, then it would double the national assets virtually overnight (Brown 2012).

Needless to say, this would make Silicon Valley and the United States aerospace establishment exceedingly wealthy, far wealthier in the case of Silicon Valley. $100 trillion divides a lot of ways. With just 1/1000 of a percent of the equity, a former scientist at NASA would become a billionaire. Obviously, this could make thousands upon thousands of Silicon Valley entrepreneurs and NASA personnel billionaires. According to Forbes, there are around 2,000 billionaires in the world today (Forbes 2014). This one entity could create 10–20 times as many billionaires as there are in the rest of the world combined from out of Silicon Valley, NASA, and the American aerospace industry. Anyone having difficulty believing that these assets could generate this much wealth need only take one hard look at the Gulf Arabs to see the tremendous wealth that energy assets can create. What is more, not only would the artificial moons deliver energy on a much larger scale than the petrochemical operations in the Persian Gulf, but also the profit margins would be higher as well.

In addition to the vast wealth that this aerospace and defense entity would create, there would be a precipitous rise in revenue to the United States treasury. If indeed there were over $5 trillion in profits, the second division could reinvest around $2 trillion back into space. This would leave roughly $3 trillion for the United States government to tax, resulting in direct tax revenues slightly in excess of $1 trillion. The rest of the profits, around $2 trillion, would be left for the fourth division to engage in investment operations. These operations would put the sovereign wealth funds of the Middle East and Asia to shame. However, if the second division were reinvesting $2 trillion into space, then this would result in a flurry of business activity on American soil that would rise significant additional revenues to the United States treasury. In addition, there would be a wave of innovation from the spillover from space that would spark a technological explosion that would cause the domestic economy to surge further. The cumulative effect of all of this could be a massive budget surplus for the United States government—perhaps in the trillions of dollars. This would mean that the United States government would not have to reform Social Security and Medicare when the baby boomers move into retirement in around 2020. Indeed, this would mean that the federal government could even increase the benefits paid out to these retirees. What is more, there would still be plenty of money left over with which to pay down the national debt.

Another important aspect about this infrastructure in space would be the ability of the company to use it as bases on orbit. The energy supply is already there for space-based radar, which would also be a necessity for the protection of the most massive of these assets. But with the energy supply, a whole range of base activities would then be enabled. Whole villages could spring up around the artificial moons.

Clearly, these assets could serve a dual function as ballistic missile defense (BMD) in addition. It would be possible to weaponize part of the energy supply to constitute directed energy weapons (DEW) in space that could shoot down missiles, especially in the boost phase. Indeed, it would likely to be technically easier to inflate a solar-pumped laser on orbit than to proceed with the conventional space-based lasers (SBL) that the pentagon has had on the drawing board since the strategic defense initiative (SDI). At least this would be the case for the small inflatables. Regardless, with hundreds of the smaller spheres, there would necessarily be a considerable amount of redundancy to have an effective missile defense constellation. What is more, as the smaller spheres could be orbited in one launch and would involve no assembly, they could be rapidly replaced in the event of war. Quite obviously with the massive DEW available, the constellation will be poised to protect itself from any number of threats. The inflatable solar power satellite could revolutionize modern warfare with its ability to neutralize intercontinental ballistic missiles (ICBMs). In fact, it is quite possible that the energy constellation could neutralize the entire missile arsenals of all nuclear powers. As such, the inflatable solar power satellite could be the greatest constant tactical factor in the history of constant tactical factors.

A Debris Service

Beyond the energy constellation, the medium of space has huge strategic significance. In the twenty-first century, the medium could develop into the most significant theater of military operations. John F. Kennedy recognized as much in 1960 when he stated "Control of space will be decided in the next decade. If the Soviets control space they can control the Earth, as in the past centuries the nations that controlled the seas dominated the continents" (Johnson 1987, 27).

In this regard, digital nations have centers of gravity that are critical to their functioning (Gray 1999). Space-based assets constitute the center of gravity of digital nations. Assets such as the global positioning system enable the digitization of national economies. The timing signal of the GPS has become ubiquitous for ATM time stamps across digital nations. Not only does the GPS perform this critical function, but also its positioning capacities have become vital for a vast array of other commercial activities as well "ranging from just-in-time logistics, international air and maritime traffic control, and the functioning of cellular telephone networks" (Sheldon 2007). To state Carl von Clausewitz's (1976) famous dictum, "one must keep the dominant characteristics of both belligerents in mind. Out of these characteristics a certain center of gravity develops, the hub of all power and movement, on which everything depends. That is the point against which all our energies should be directed" (720).

Hays (2009) has addressed the critical nature of space-based assets as global utilities. According to Hays (2009), the global information grid could not operate effectively without space-based assets. As such, commercial space activities have come to be seen as global utilities. The capabilities these assets provide include "communication, environmental, position, image, location, timing, or other vital technical service or data to global users" (Hays 2009, p. 184). Thus, space-based global utilities have become as important to modern society as water and electricity services (Hays 2009).

Therefore, with so much of global society already dependent on space-based assets (a dependency that will only grow exponentially when the globe's energy supply is constituted on orbit), the Joint Stock Company would naturally constitute a service to remove orbital debris to protect these critical assets. Thus, the second feature of the first division would be a service to remove space debris. Regarding the issue of space debris, the 2007 Chinese direct ascent anti-satellite (ASAT) test created a tremendous amount of orbital debris in low Earth orbit (LEO). In fact, the International Space Station (ISS) has had to adjust its orbit numerous times to avoid hitting the space debris that this act generated. More recently, the American Congress has looked into the issue of space debris.

What they found is that "decades of human space flight—primarily U.S. and Russian space activities—have littered the Earth's orbit with debris. NASA defines orbital debris as 'all man-made objects in orbit about the Earth which no longer serve a useful purpose.' Examples include derelict spacecraft, abandoned space launch vehicle stages, mission-related debris, and fragments created as a result of explosions or collisions" (Hildreth and Arnold 2014, p. 2). For many decades, now the United States has taken the international lead in monitoring all of the trackable objects in Earth orbit. The United States Air Force (USAF) has traditionally operated the debris observing surveillance equipment for the United States. In this regard, "the US Space Surveillance Network was the leading space object tracking system in the world and catalogued objects as small as about 10 cm (softball size) in LEO and as small as 1 m in Geosynchronous Orbit. In the past, the Space Surveillance Network tracked more than 23,000 objects 10 cm in diameter or larger in orbit around the Earth. Of those, only about 1,100 (5 %) were active satellites. The rest was orbital debris. In addition to the debris tracked by the Space Surveillance Network, there were hundreds of thousands of pieces of debris smaller than 10 cm, which were considered too small to track or catalogue, but were still capable of damaging satellites and the International Space Station" (Hildreth and Arnold 2014, p. 2). But while the USAF has done a good job tracking debris with the systems they have had in the past, the first division of the Joint Stock Company will be poised to move out with even more technically sophisticated systems for tracking objects in space. Indeed, the USAF already has a new and more sophisticated space fence under development. "The new Space Fence will track about 200,000 objects and make 1.5 million observations per day, about 10 times the number made by previous radars such as the AFSSS. Air Force leaders have estimated that the actual number of objects orbiting Earth is closer to 500,000. The new system would have a maximum coverage area of 40,000 km, whereas the previous AFSSS covered 22,000 km maximum" (Gruss 2015). Perhaps the company would acquire the same technology when it becomes available in the near term while devoting even more resources to the development of even more advanced space object tracking systems over the long term.

In addition to such tracking technologies, the company will have an incredibly sophisticated space situational awareness (SSA) architecture as well. The company's SSA will certainly make use of the most advanced nanosatellites for the effort. In fact, the energy constellation will be constantly surrounded by co-orbiting satellites. All of the monitoring activities occurring in the artificial moons' orbits will be tremendous to say the least.

Beyond the USAF's efforts to track debris, NASA pays close attention to the problem as well. In 2014 alone, NASA reports that until September at least "seven small suspected and confirmed breakups have occurred in low Earth orbit since late March. The first was Cosmos 1867, a Plazma-A-class spacecraft launched by the former Soviet Union to test a new, advanced nuclear power supply. Cosmos 1867 is a sister to Cosmos 1818, which created a similar debris cloud in July 2008. As with Cosmos 1818, the cause of the breakup is unknown, although six objects identified as 'coolant' were added to the SSN catalog of orbiting objects. It is suspected that the debris are leaked sodium potassium coolant released either through a hypervelocity impact of a small particle or some other breach in a coolant tube through thermal cycling" (NASA 2014, p. 1). NASA believes that these breakups on orbit were due to hypervelocity impacts of small meteoroids or other debris (NASA 2014).

Space debris is a critical issue that highlights the important privatization dynamic that the Joint Stock Company creates. As with the arrangement for LEO space travel with NASA, the American government may need a similar arrangement with a private entity to remove debris because it may not have the financial wherewithal to do the job in all of its totality all by itself. Some have already proposed to the Congress that it begin exploring the budgetary costs of debris removal and have suggested that these costs could be as significant as any major space program (Hildreth and Arnold 2014). Such difficult budget realities may necessitate that the Congress, the Obama administration, and future American administrations take the same approach to debris removal as was taken with the industry–government partnership for LEO space travel. As such, the Joint Stock Company will be poised to fulfill this vital role for the United States government as well as for the rest of the world.

There is definitely a strong business case to be made for the removal of space debris even if the most threatening debris is in LEO where there are few commercial satellites. Indeed, "orbital debris in GEO, which 'moves in an enormous doughnut shaped ring around the equator as the gravitational forces of the Sun, Moon and Earth pull on the objects,' is 'not naturally removed from orbit by atmospheric drag,' and thus is 'estimated to last anywhere from a million to 10 million years.' Moreover, it has been estimated that collision risk in GEO 'is not uniform by longitude,' but instead is 'seven times greater in regions centered around the so-called 'geopotential wells' which exert a gravity pull on drifting satellites and other debris.' According to the insurer Swiss Re, there are operating satellites worth 'hundreds of millions of dollars' that are 'in or near these locations'" (Nelson 2014).

What is more, any rocket passing through LEO to loft commercial satellites into GEO could get struck by space debris causing billion dollar investments to implode. Additionally, if collisional cascading occurs, no commercial operators will be able to access space. Collisional cascading occurs when debris strikes other debris, or other satellites, which then creates more debris, which then strikes more and more debris and satellites until so much debris is created that it encases the Earth (Brown 2012). If collisional cascading occurs, the medium of space will be barred for all participants, including commercial entities from all spacefaring nations. And even in GEO the growth of space junk may eventually become so large in the twenty-first century that orbital slots will tighten necessitating its removal to make room for new satellites.

The question of space debris has heretofore posed a quandary for American strategists. Nations from around the world from Brazil to Iran to China to North Korea are all developing space programs varying in their level of sophistication. Added to this mix are a multitude of private entities, such as Space Exploration Technologies (Space X), which are now multiplying in the medium. With the exponential growth of space activity that is set to ensue in the twenty-first century, it is inevitable that an unmanageable amount of space debris will be created in Earth's orbit. If too much debris is created, collisional cascading is sure to result. It is therefore imperative that the United States develop the capability to remove debris before it is too late.

To this end, the Joint Stock Company could undertake to modify many prototype ASAT weapons to remove debris. As debrises are also satellites of the Earth, these weapons could serve a dual purpose. Indeed, ground-based lasers and space-based lasers could be utilized to ablate debris in LEO and lower its altitude. Co-orbiting assets could be deployed in higher orbits to rendezvous with defunct satellites and move them to graveyard orbits before they collide with other satellites and create debris (Brown 2012).

While the use of lasers to ablate debris is the preferred method for its removal because they would be more cost effective and would face less technical challenges than other methods (at least the solar-pumped lasers from the inflatable solar power satellites would present this technical–economic dynamic), DARPA has nevertheless already begun exploring a variety of other exotic techniques to remove large debris. DARPA has found that "the removal of large objects generally employs advanced rendezvous and proximity operations and sophisticated grappling techniques. Various methods of capturing large objects have been proposed involving a net, inflatable longeron, tethered harpoon, articulated tether/lasso, and an electrostatic/adhesive blanket. Some solutions attached or used an active thrust device, while others made use of natural forces found in the space environment to impart a force on the debris to relocate it" (Hildreth and Arnold 2014, p. 11).

Already DARPA has been exploring cooperation between the public and private sectors for some of its activities in space. "Under a demonstration project called Phoenix, DARPA is teaming up with the private sector to harvest and 'repurpose' still functional components of nonworking satellites in GEO to create new space systems at greatly reduced cost. Beginning in 2016, the project proposes to attach nanosatellites to parts of retired US government and commercial satellites, making the debris a resource. In a process called, 'cellularization,' nanospacecraft separately carrying out functions such as power, communications, and attitude control would be launched into orbit as secondary payloads" (Anzaldua et al. 2014).

At the same time, interests from the private sector are emerging with plans to remove, recycle, "or reuse large debris in LEO. For example, three companies—Star Technology and Research, Inc., Tether Applications, Inc., and Electrodynamic Technologies, LLC—have been developing a technology called ElectroDynamic Debris Eliminator (EDDE), wherein a long conductor is energized using solar energy to thrust against the Earth's magnetic field. Operating without propellant, EDDE can repeatedly change its altitude by hundreds of kilometers per day and its orbital plane by degrees per day" (Anzaldua et al. 2014). With such developments in the offing, it would now be natural for the Joint Stock Company to fully combine the expertise that government agencies can bring with the dynamism characteristic of these nascent industry efforts into one centralized effort with the aim of removing debris to protect the energy infrastructure as well as all other critical assets in the medium. In this regard, the Joint Stock Company would recruit the expert personnel from the government agencies in question as well as acquire the smaller debris removing private enterprises that were demonstrating a critical aptitude for these pursuits. With everyone combined into one centralized effort, the Joint Stock Company will be fully poised to move out with the removal of space debris in Earth orbit.

In any event, regardless of the technique utilized to remove the debris, the assets employed on orbit in this capacity could be dually used to target enemy assets in addition. Therefore, not only would this ASAT fleet deployed under the auspices of removing debris protect the energy infrastructure and other critical assets in space, but it could also pursue other financial opportunities in the medium as well. Bounties could be placed on debris in the path of other space participants, or the first division could charge other space participants a flat fee to patrol space lanes and keep them clear of debris (Brown 2012). Perhaps the entity would even go so far as to stealthily fowl the orbits of satellites from other nations that were not under the aegis of the Joint Stock Company in space (Brown 2012). This activity would not necessarily entail a protection racket, but rather it would be similar in scope and effect as the strategy that the British Empire pursued on the high seas to protect vessels from nations with which it had trade advantages and to prey on vessels from nations with which there was trade competition. In any event, in this manner, the first division could exercise a considerable measure of control in the commons of space.

Indeed, the debris service would necessarily constitute a space "fleet in being." Its very presence would serve to deter nefarious activities that other space participants may present (Brown 2012). This would create a very favorable strategic dynamic for the Company in space.

Of significant strategic importance would be the state of readiness that the first division's forces would be in from their constant debris removal operations. These operations would prepare the first division quite well for war if it did break out in the medium. Not only would the first division be better prepared to target enemy assets in space, but it would be more equipped to deal with debris if war led to the creation of a considerable amount of debris (Brown 2012).

Another critical security dynamic related to the debris fleet would be the ability of the company to proliferate weapons capable of targeting the Earth under the auspices of the debris removal business. These weapons could range from the "Rods from God" to orbital bombardment weapons to fractional orbiting weapons to ICBMs. Such systems may prove necessary if the company is to deter or thwart potential attacks against the energy constellation.

Additionally, the debris service could perform a whole suite of services for the satellites of other space participants. These could range from maintenance to servicing to upgrades, all of which could extend the life of satellites of other space participants. These activities would increase the intelligence capabilities of the debris service tremendously and would allow the fleet to keep better track of the space environment (Brown 2012).

Intelligence

Intelligence operations would span the whole of the organization. There would be stovepipes for not only the first division, but also for the business activities of the third division and the financial activities of the fourth division as well. It is likely that the intelligence activities of the third division would be engaged in a significant amount of industrial espionage, while the intelligence activities of the fourth division would seek to ensure that the most comprehensive and highest quality of information was at the disposal of its trading operations. The intelligence activities of the first division would include security matters of course. It is anticipated that a significant amount of former CIA and DIA officials, analysts, and agents would be employed by the first division. The first division could easily afford to pay them far higher salaries than they were making at government agencies. In addition, they could become millionaires and billionaires when they received the company's stock options. The company's operations would obviously be far more lucrative career moves for them.

Obviously, the company would possess the most sophisticated intelligence assets in space, including photo, signal, and electronic intelligence satellites. In addition to imagery collected for security purposes, perhaps the company would also create a service to share more precise observations of the Earth with global society. Clearly, the first division would be very busy intercepting the signals of states, entities, and/or other actors that posed a risk to the company. Regarding the elint capabilities of the first division Johnson (1987) illustrates how these help determine enemy orders of battle:

> A kindred class of satellites monitor the Earth, not in the optical portion of the electromagnetic spectrum but through the radio and radar atmospheric windows. These electronic intelligence (elint) gathering satellites listen patiently to the radio and radar emissions of ground, air and sea emitters. For example, the detection and location of emissions of the

> search and tracking radars of mobile air defense units provide valuable assistance to the planning of tactical air strikes. Likewise, the interception of communications may disclose not only the contents of the messages, but also the locations and identities of the communicators. Together these types of information permit the construction of an electronic order of battle (EOB) which would influence the conduct of an engagement (60).

Such elint assets will no doubt enable the first division to detect preparations for any assault that foes may make against the company's critical assets in the medium, most notably against the artificial moons. The first division of course would also be in possession of early warning satellites for these purposes as well. Perhaps the first division will also constitute and operate a navigation constellation to augment the services that the GPS provides. Obviously, the company will develop its own set of totally secure communications satellites for its operations as well.

Cyber

Cyber operations would span the whole of the organization as would be the case with the company's intelligence operations. The cyber capabilities of the Joint Stock Company would be more than considerable due to the company's ability to compensate cyber personnel at the highest possible levels. Indeed, the first division's security operations could employ the entire NSA and would have the resources to more than double their salaries. Critically, the company could give the deftest cyber personnel enormous stock options. In this way, the company could attract the top talent in cyberspace and make thousands of cyber personnel billionaires. In fact, the company would likely be able to compensate the top cyber personnel far more than if these personnel were to take their own software companies public on stock exchanges around the world. Certainly, the company could make far more of them wealthy than would otherwise be the case at a typical software company in the IT industry. Indeed, the company could create thousands more IT billionaires than there are in all of Silicon Valley combined. What is more, the company would without doubt constitute a constellation of information assets on orbit to provide internet access to every inch of the globe.

A glance at the first division sees a global energy monopoly emerging in orbit with the medium being rapidly secured by a debris fleet. The picture then contains significant intelligence and cyber operations, all with unlimited resources.

Chapter 3
The Second Division—Space Colonization

Once the energy operation was secure and the company had established a global energy monopoly, the company would have tremendous resources with which to reinvest in space—trillions of dollars. This would continue the development of the company's core business. As such, the company would have the financial wherewithal to create space civilization. Undergirding space civilization would be a fledgling space economy. In this regard, while the energy infrastructure would be the foundation of space civilization, a burgeoning trade in natural resources extracted from around the solar system would become the means with which to build the economic artifice. Therefore, the first objective of the second division would be to explore for natural resources in the planets and asteroids throughout the solar system. Indeed, some have speculated that there may be as much as a trillion dollars worth of water ice and minerals in a number of near-Earth asteroids (Garretson 2013). Obviously, far more money is waiting to be unlocked in the planets.

Once the second division was able to locate and reach the resources, it must extract them. Once the resources are extracted, they must be studied and analyzed. Once they are adequately analyzed, they must be processed en masse. After they are processed, they must be fabricated or put to use in other ways. Clearly, this process would be a massive undertaking in space. Therefore, in order to support this undertaking, it would be necessary for the second division to colonize the inner solar system. In this way, space civilization would arise upon the foundation of the Solar Moons. Clearly, there would be a very brisk trade in space.

In Space Manufacturing

To start the second division could create an in-space manufacturing capability in Earth orbit. As has been mentioned, villages would spring up around the artificial moons. Naturally, they would possess the capability to support the energy

T. Brown, *The Digital Galactic Complex*, SpringerBriefs in Political Science 30,
DOI 10.1007/978-3-319-22386-5_3

infrastructure, while simultaneously producing products in micro-gravity for consumption on the Earth. However, these facilities could also support deep space exploration by serving as depots and supplying missions. Perhaps at some point in the distant future these facilities would also fashion and assemble very large spacecraft. A ring around the globe could easily arise in LEO as manufacturing develops in space. The advent of 3D printing certainly increases the capacity to perform these activities.

Regarding manufacturing on orbit, the medium of space has special characteristics that make it uniquely suited to certain forms of extremely sophisticated manufacturing. In particular, the micro-gravity in Earth's orbit significantly enhances all efforts at materials processing (McDonnell Douglas Astronautics Company 1977). The special properties of Earth orbit include "hard vacuum, low vibration, wide-angle view, and a complete isolation from the earth's biosphere…in fact, in-space pressure levels of one-trillionth of an atmosphere are relatively easy to obtain" (Rockwell International Space Division 1978, p. 1). These attributes make the environment very suitable for all forms of space industrialization, which may be considered as the leveraging of new technologies in the unique environment of the medium "for the social and economic benefit" of terrestrial society (Rockwell International Space Division 1978, p. 1).

Furthermore, "vacuum levels ranging from 10^{-2} to 10^{-10} atmospheres have already been used in a number of practical ways. These include food processing and preservation (including freeze drying and refrigeration), metal distillation, X-ray devices, TV picture tubes, thin film deposition, and the manufacture of vacuum diodes and solid state electronic devices" (Rockwell International Space Division 1978, p. 1). "Precise g-level control is important in medical and chemical centrifuges, in crystal growth, electrophoretic separation, solidification and purification processes, and in the construction of extremely lightweight orbiting structures such as multibeam antennas" (Rockwell International Space Division 1978, p. 1).

Indeed, the medium of space affords fantastic opportunities for the development of sophisticated new production techniques that maximize the absence of terrestrial conditions such "as natural convection and sedimentation" (McDonnel Douglas Astronautics Company 1978, p. 1). By fully exploiting these special conditions in the medium, the second division will not only be able to develop more technologically sophisticated goods than could be produced on the Earth, but it will also be able to create entirely new revolutionary technologies. While certain attributes of the medium can be replicated on the Earth, such as high vacuum and radiation, the most vital attribute, weightlessness, can only be had for a very short time (McDonnell Douglas Astronautics Company 1978). The micro-gravity of orbit allows molten materials to be suspended naturally with minimal equipment. Such dynamics necessarily allow for the complete elimination of contaminants (McDonnell Douglas Astronautics Company 1978). But what makes the medium the most critical for in-space production is the ability to escape gravity-induced convection. In this regard, "the thermal gradients in fluids" cause convection currents, which ultimately result in a certain amount of difference from product to product that are mass produced on the Earth (McDonnell Douglas Astronautics

Company 1978, p. 1). "Having escaped the problems posed by these currents, space manufacturers will be able to grow crystals of great purity with highly controllable characteristics; they will find it much easier to mix and homogenize liquids, to cast metals, and to separate and purify the elements of mixtures" (McDonnell Douglas Astronautics Company 1978, p. 1).

The biggest economic challenge for production operations on orbit is that the materials that are produced there might be too expensive for terrestrial markets, or another material might be developed on the Earth which would call into question the value of space-produced products such as those made with silicon. Therefore, the biggest factor contributing to the success of products produced on orbit is the capacity to charge a high enough price for them in terrestrial markets. This dynamic will ensure a decent rate of return on investment. For products made with silicon that are produced on orbit, this would necessitate a price that was both above its material cost and reflected the technical improvements of the product (McDonnell Douglas Astronautics Company 1977). However, comparing space-produced goods to those on the Earth is not the right comparison. All of the additional technical improvements that go into space-produced material must be calculated and determined to assess their economic viability (McDonnell Douglas Astronautics Company 1977).

In any case, the second division will be totally confident that products produced in space will possess high values. It will fully understand "that the benefits of processing in space will be" significantly greater than those on the Earth—that the medium will allow for the production of more sophisticated products than those on the Earth and even that the medium will allow for the production of products that cannot be produced on the Earth (McDonnell Douglas Astronautics Company 1978, p. 1). As such, the company will be totally prepared to begin investing in-space production processes even if they are largely untried at the present (McDonnell Douglas Astronautics Company 1978). In this regard, the company will have complete confidence that its production processes can be created quickly and at reasonable cost, and that markets in space and on Earth exist that will allow products to be sold at the necessary price levels to ensure a solid rate of return on these investments. What is more, it is highly unlikely, probably even impossible, that a new product could be produced on the Earth that could substitute for the variety of products made in space and capture off the Earth markets. In addition, the plausibility that a breakthrough in terrestrially produced technology will allow for competitive ground production is low as well (McDonnell Douglas Astronautics Company 1978).

Perhaps the most critical dynamic about the company's production operations in orbit that will make them extremely viable and competitive in terrestrial markets will be the massive amounts of transportation activities that will take place between Earth and orbit. Indeed, the company will be launching incessantly into space. With such high launch rates, the cost of each launch and rocket will plummet drastically, which will necessarily mean that transportation costs for products produced in orbit will be very low making these products much more competitive in terrestrial markets.

The United States civil space program has already experimented with a number of materials that are used in incredibly sophisticated technologies. In this regard, there were a number of experiments on Skylab for crystals to be utilized in semiconductors. A few of the most promising experiments involved "uniform distribution of dopant elements and crystal facets which were flat within a few hundred Angstroms" (McDonnell Douglas Astronautics Company 1977, p. 1). "Single crystal silicon for use in integrated circuits requires extremely high quality material in terms of purity and structure. The damaging effect of defects increases with circuit size and is greatest for large scale integrated circuits (LSI)" (McDonnell Douglas Astronautics Company 1977, p. 2). This is but one example of why manufacturing in space is such a fantastic opportunity. Once again, the unique characteristics of the medium will necessarily allow for large-scale production operations to take place that will eliminate defects in sophisticated hardware.

In studies undertaken by elements of the American aerospace industry in the late 1970s, it was found that plant and equipment were the primary costs for production with silicon in space and that transportation by the shuttle would have only been a minor cost (McDonnell Douglas Astronautics Company 1977). "Return on investment for only one plant" would have been extremely low due to the costs of "design, development, test, and engineering" being "equal to the plant cost" (McDonnell Douglas Astronautics Company 1977, p. 8). However, if the company were to achieve a large measure of scale with its production operations in space, then these costs could be spread out over a number of plants on orbit thereby decreasing the unit cost for each plant and thus yielding a much larger rate of return on investment.

Another economic area for space production operations involves the development of pharmaceutical products. In this regard, it might be possible to automate facilities on orbit to separate chemicals for pharmaceutical production. Indeed, in the 1980s, McDonnell Douglas even thought it might be possible to produce a cure for diabetes through the process of electrophoresis in orbit (Yardley 1983).

What is more, the advances in 3D printing are opening up further opportunities for production operations on orbit. "Using 3D printers as production tools has become known in industry as 'additive' manufacturing (as opposed to the old, 'subtractive' business of cutting, drilling and bashing metal). The additive process requires less raw material and, because software drives 3D printers, each item can be made differently without costly retooling" (The Economist 2011).

3D printing might seem to some to be somewhat strange. However, the process is largely similar to traditional printing that involves the transmission of digital files from a computer to a printer. What differentiates the two processes is that the "ink in a 3D printer is a material which is deposited in successive, thin layers until a solid object emerges" (The Economist 2011). The layers are determined by software that calculates all the levels digitally with a computer-enhanced design. The digital details of each level of the object are transmitted to the device to build the levels in succession. The object can then be assembled in a variety of ways. "Powder can be spread onto a tray and then solidified in the required pattern with a squirt of a liquid binder or by sintering it with a laser or an electron beam. Some machines deposit filaments of molten plastic. However it is achieved, after each layer is complete the build tray is lowered by a fraction of a millimetre and the next layer is added" (The Economist 2011).

"Today, 3D printers have evolved to make a variety of objects using a laser or extruder (the material output part of the printer, best described as a futuristic hot glue gun) that move along an X, Y and Z axis to build an object in three dimensions, layer by layer, sometimes only microns thick at a time, depending on the desired resolution of the object. This method eliminates a lot of wasted materials, as any leftover powdered substrate can be immediately used on another project, alleviating the need for injection molding, setup costs, cutting, sanding, drilling and having scraps of material left over, as is common with traditional manufacturing methods. The most impressive part: economies of scale cease to be an issue as costs for single parts become standardized in relation to the costs of the material being used" (Hart 2012).

The United States military has been evaluating 3D printing as a more economical and sophisticated way to resupply spare parts for its aircraft at sea and in the main theaters of combat. For similar reasons, NASA has already been looking at the use of 3D printers for its operations in space. Because the cost of launching materials into space is so high, NASA has been investigating the ease at which it could produce items for its missions in space with 3D printers (Hart 2012). Resupply has always been a challenge for NASA's missions. But with 3D printing, it may be possible to create parts remotely which may reduce NASA's need to store parts in space (Hart 2012).

Beyond 3D printing operations in Earth orbit, another example may be for operations on the Moon. Silicate minerals are in large supply on the Moon. With a 3D printer, all that may be required is a binding agent to fashion products from these minerals (Hart 2012). As NASA has already been exploring the dynamics of 3D printing in the medium, the second division would no doubt be poised to exploit these dynamics to the fullest.

Operations on Mars are yet another example. As the company will most definitely seek to build habitation infrastructure on the Red planet, the company can transform the cargo being shipped to Mars "from bulky parts to raw bulk material" (Hart 2012). The personnel of the second division can simply transmit a signal to a 3D printer on Mars through the recesses of space with commands to print dozens of more 3D printers. These printers could easily be of assorted types necessary to build infrastructure, equipment, and a variety of vehicles. Indeed, much of the raw materials required to develop these systems may already be located on Mars, allowing the second division to further reduce supplies to be shipped to the planet (Hart 2012).

Finally, these manufacturing facilities on orbit could also support a thriving tourist industry. They could easily host hotels on orbit as well as recreational activities. Clearly, these facilities would be tremendously beneficial for astronauts passing through Earth orbit as well. As such, these villages would necessarily serve as oases in Earth orbit.

The Moon

Beyond operations in Earth's orbit, the Moon looms as vital strategic real estate. The Moon would necessarily serve as a staging ground for missions across the solar system in search of natural resources. The Moon is extremely valuable in this regard not only for its location but also because the resources located there could be utilized in situ to support the exploration missions in addition to the oases in Earth's orbit. Therefore, it would be critical that the second division develops a base of mining and manufacturing on the Moon to support the burgeoning space economy. Significantly, the company would easily have the financial wherewithal to systematically surmount the major technical obstacles to a lunar colony. Obviously, the enormity of the traffic between the Moon and Earth would dramatically reduce the cost of transportation. This dynamic would be vitally important for the fledgling space economy even if the second division were in possession of unlimited resources.

Lunar habitation has seemed like science fiction for a long time. Ever since authors like Jules Verne were writing about travel to the Moon outposts, there have captured many imaginations. "One of the earliest practical proposals was the US Army's 1959 design for a nuclear powered fortress, built to establish a military presence on the Moon before the Soviet Union could do the same" (Tate 2013). "A 1961 US Air Force plan called for a 21-man underground lunar base, to be built by 1968" (Tate 2013). But while such infrastructure on the Moon could possibly have been feasible in that era if the economic resources and the political will had materialized, habitats on the Moon are becoming increasingly viable today. Indeed, with the advent of advanced robotics technologies and the possibility of integrating their operations with human activities, it has become increasingly possible to create infrastructure on the Moon which could support human life. In this regard, the robotics technologies could precede human landings on the Moon and prepare sites for the astronauts. 3D printers could also be utilized to create parts for infrastructure from the lunar regolith (Tate 2013).

In any event, a lunar outpost must support the variety of needs that human life requires. Such support must either come from supplies transported from Earth or by utilizing resources in situ. Fortunately, the water requirements for life on the Moon are lower than those on the Earth. "On Earth, the daily life-support requirements for one person are:

- Oxygen: 1.85 lbs (0.84 kg)
- Drinking water: 2.64 gallons (10 l)
- Dried food: 3.9 lbs (1.77 kg)
- Water for food: 1.06 gallons (4 l)" (Tate 2013).

"In space however, the water requirements are lower:

- Oxygen: 1.85 lbs (0.84 kg)
- Drinking water: 0.43 gallons (1.6 l)
- Dried food: 3.9 lbs (1.77 kg)
- Water for food: 0.21 gallons (0.8 l)" (Tate 2013).

What is more, the fact that the basic requirements for life can easily be had from the Moon's soil makes human habitation there increasingly possible. In this regard, air and water can be separated from the soil on the Moon. Additionally, materials necessary for infrastructure and fuel for spacecraft could also be easily produced (Tate 2013).

Perhaps the most attractive location for initial settlements on the Moon would be at the lunar poles. While these locations would be in permanently shadowed deep craters and would be exceedingly cold, they do in fact contain the largest amounts of water ice that can be found on the Moon (Tate 2013). "Scientists think that water might be responsible for the fluffy dirt at the Moon's poles. Small particles of water frost moving in and out of grains of dirt may result in more holes between the grains, giving it a porous texture" (Moskowitz 2012). Prior studies concerning the possible amounts of water on the Moon found that as much as 0.5 percent of water could be found at more central geographical locations that receive more sunlight. But more recent studies have discovered that water might be found at up to 2 percent in the soil in the regions in permanent shadow (Moskowitz 2012). Such supplies of water could probably be extracted with relative ease (Tate 2013).

Furthermore, the location of the Moon relative to the Earth makes it a very attractive destination for the second division. The Moon will necessarily serve as a staging ground for the second division's missions across the solar system. With all the activity that would be occurring on the Moon, the cost of transportation would surely decrease dramatically. In addition, much infrastructure would develop in cislunar space to support operations on the Moon. "By creating a lunar cycler, a station at L-1, an orbital fuel depot, in situ utilization of lunar oxygen and possibly lunar water, there could be a vibrant community on the Moon" (Dinkin 2004).

The proximity of the Moon also holds out the possibility of performing rescue operations to the Moon and in the Moon's vicinity from the Earth. As with NASA's shuttle operations, the ease at which the second division could develop standby operations for lunar emergencies would not require significantly increased capabilities from those that would already be in existence. The hardware that the second division would be in possession of would easily allow for all types of contingencies in the event of some unforeseen disaster on the Moon (Dinkin 2004).

"The proximity to Earth allows for just-in-time planning. With Earth only a few days away, a regular resupply mission can have last minute changes to its manifest. That means that fewer spares need to be kept on hand to assure the same level of safety" that would otherwise be required for other missions throughout the solar system (Dinkin 2004).

The proximity of the Earth to the Moon will also allow humans to operate robotics technologies from the Earth to augment operations on the Moon as opposed to the necessity of humans performing all robotics operations on the Moon. This dynamic will enable the second division to further increase its ability to utilize capital equipment in lunar operations (Dinkin 2004). Such a scenario will make the process of colonization much easier at this location than it will be elsewhere in the inner solar system.

Indeed, by establishing a colony on the Moon straightaway as the next destination for the second division, there will be many valuable lessons to be learned with this effort that will increase the probabilities of success in all other colonization efforts elsewhere in the inner solar system (Dinkin 2004). It seems as though the Obama administration's space policy, which is concentrated on asteroids and then Mars, may miss such lessons by skipping the Moon altogether. In this regard, after production operations in Earth orbit, the Moon is the first logical destination for the second division. The Moon could then become a staging ground for missions to Mars and elsewhere across the solar system.

Furthermore, once the second division had invested humans on the Moon, many options would then be opened to conduct missions to near-Earth asteroids. All of the infrastructure on and about the Moon could be utilized for missions to rendezvous with asteroids and extract water ice and minerals, and even to bring some of the asteroids back to cislunar orbit. Such operations could further supply the second division's operations on the Moon and elsewhere into deeper space.

Obviously, with all of the resources that could be exploited on the Moon, and with the Moons pivotal location as a staging ground, the colony on the Moon would easily become close to self-sufficient. In this regard, the Moon may also become a very attractive tourist destination as well. A vibrant tourist industry could easily arise on the Moon at the same time that it was developing in Earth orbit. Furthermore, the drastic reductions in the cost of transportation to the Moon will enable a critical mass of tourists to visit the Moon for relatively reasonable sums of money. Indeed, as oxygen and water are more easily obtained on the Moon as opposed to orbit, the location may even become a cheaper destination for tourists than orbit (Dinkin 2004). In any event, the lunar colony could probably host larger crowds than the infrastructure in orbit.

Some might even consider the Moon to be a more exotic destination than Earth's orbit, which could mean a larger tourism industry there. Should a strong tourist industry arise on the Moon it will lead to an economic dynamic requiring less subsidy from the company to support the colony. Such a dynamic could probably be obtained relatively quickly once the company starts its operations there. While the colony would still require imports from Earth, as all countries require imports from elsewhere on the Earth, self-sufficiency in all goods will not be required for the colony to succeed (Dinkin 2004).

The lunar colony would also afford a fantastic area for astronomy (Dinkin 2004). The facilities for such activities would be state of the art as well. It would be a veritable playground for astronomers and astrophysicists.

What is more, "space skills will be valuable and firms and people with experience on the Moon will be well able to help develop cislunar and martian systems. Radiation management experience, artificial gravity creation technology, operation and maintenance, flywheel, maglev, and mass driver technologies are all likely to be developed on the Moon and useful in future efforts" (Dinkin 2004).

With all of the resources that can be utilized in situ on the Moon, with its critical capability to serve as a base of operations for exploration missions, and with all the outstanding tourism potential, a colony on the Moon would be an extremely lucrative operation for the Joint Stock Company. It would obviously be pivotal for the company's efforts to colonize the inner solar system. Indeed, the second division could leverage the development of a lunar colony as a huge pillar in the foundation of space civilization.

Comets and Asteroids

Moving further out into space from the Moon and cislunar orbit, the second division would next seek to exploit the tremendous resources that comets and asteroids may hold. There is much speculation about the mineral wealth in some near-Earth asteroids—some have suggested that there may be as much as a trillion dollars worth of water ice and minerals in some of these objects (Garretson 2013). Evidently with sophisticated telescopes, it may be possible to determine what the comets and asteroids are composed of based on the light from the Sun that is reflected off of them. In any event, the comets and asteroids are the next logical step for the second division's operations into deeper space. In this regard, JPL (2014) defines these objects in the following way:

> Near-Earth Objects (NEOs) are comets and asteroids that have been nudged by the gravitational attraction of nearby planets into orbits that allow them to enter the Earth's neighborhood. Composed mostly of water ice with embedded dust particles, comets originally formed in the cold outer planetary system while most of the rocky asteroids formed in the warmer inner solar system between the orbits of Mars and Jupiter. The scientific interest in comets and asteroids is due largely to their status as the relatively unchanged remnant debris from the solar system formation process some 4.6 billion years ago. The giant outer planets (Jupiter, Saturn, Uranus, and Neptune) formed from an agglomeration of billions of comets and the left over bits and pieces from this formation process are the comets we see today. Likewise, today's asteroids are the bits and pieces left over from the initial agglomeration of the inner planets that include Mercury, Venus, Earth, and Mars.

Additionally, JPL (2014) states that

> As the primitive, leftover building blocks of the solar system formation process, comets and asteroids offer clues to the chemical mixture from which the planets formed some 4.6 billion years ago. If we wish to know the composition of the primordial mixture from which the planets formed, then we must determine the chemical constituents of the leftover debris from this formation process—the comets and asteroids.

While some near-Earth asteroids may pose significant danger to life on Earth because they may pass close to our planet and could cause tremendous devastation if in fact there was impact, these objects, however, could be very lucrative for the second division's operations because their close proximity makes their mineral wealth more easily accessible (JPL 2014). Indeed, the second division would obviously be poised with the capabilities necessary for planetary defense. Regardless, although economies would not allow for the exploited minerals of these objects to be imported to the Earth, the raw materials that they contain could probably be employed for the development of infrastructure in space and to fuel the rockets that would be necessary for the colonization of the inner solar system. What is more, some maintain that the mineral deposits in the asteroid belt between Mars and Jupiter at present amount to at least $100 billion for each individual on the Earth (JPL 2014).

Both the comets and asteroids are extremely attractive opportunities for the second division. The "water and carbon-based molecules" that life depends on are abundant in the comets (JPL 2014). Furthermore, the ice on comets can easily be broken down into hydrogen and oxygen, the two elements necessary for rocket fuel. The asteroids, however, have abundant minerals necessary to create infrastructure in space and hold vast opportunities in this regard (JPL 2014). Therefore, when the second division quickly undertakes the colonization of the inner solar system, the "metals and minerals found on asteroids will provide the raw materials for space structures and comets will become the watering holes and gas stations for interplanetary spacecraft" (JPL 2014).

The opportunity that asteroids present in space is so great that already a few companies have been created to exploit them. These include Planetary Resources and Deep Space Industries. Such companies would be unlikely to challenge the Joint Stock Company's dominance in these markets because the resources that the solar power satellites would afford would allow the company to conduct operations on such a scale that it would dominate these smaller ventures. Furthermore, if conflict were to arise in space over competition for these asteroids, the Joint Stock Company would likely destroy these smaller companies due to the tremendous abilities of the security personnel that were working at the Joint Stock Company.

In any event, the mission of Planetary Resources is to address the resource scarcity presently facing the Earth and beyond (Planetary Resources 2014). Likewise, the second division would be poised to fully address this important issue as well. In this regard, the second division would also be "focused on developing the tools to grow the economy beyond Earth, including: radiation tolerant embedded systems, deep space communications and remote sensing capabilities, autonomous navigation, and integrated propulsion on the smallest, most cost-effective platforms possible" (Planetary Resources 2014).

What is more, the tremendous resources available in vast amounts on asteroids could easily be employed to make space transportation much easier "from orbit raises to station keeping to Lunar landings or even expeditions to the outer planets" (Planetary Resources 2014). As such, the main objective for the second division regarding asteroids is to exploit them to the fullest to create the infrastructure necessary for the colonization of the inner solar system.

Asteroids, which again are basically rocks in space that consist of metals and ice, vary greatly in size. They can be only a few feet in diameter or with cross sections of up to 975 km in the case of Ceres, the biggest. The type of light they reflect is how they are classified, which is "according to their spectral type" (Tate 2013).

Planetary Resources' telescopes, the Arkyd-100, will evidently be able to ascertain the composition and market value of asteroids through such spectral observations (Tate 2013). The second division will also possess the capability to make similar determinations by the type of light that the asteroids will be reflecting with its extremely sophisticated telescopic capabilities. Then the second division will conduct a variety of missions to these objects. The second division will both swarm these asteroids with small spacecraft, but perhaps it will also conduct human missions to the most promising asteroids. In the course of these operations, the second division may capture some of the asteroids and bring them back to cislunar space or Earth's orbit similar to what NASA is proposing to do under the Obama administration.

Deep Space Industries is also in the asteroid game. This company has plans to utilize the tremendous mineral wealth of asteroids for robotic production in Earth's orbit with 3D printers (Tate 2013). Obviously, the second division would also utilize its aforementioned 3D printing capabilities for similar pursuits. Deep Space Industries is planning to use cubesats for reconnaissance missions to the vicinity of asteroids. Then they are planning to send larger spacecraft to obtain asteroid samples or even bring back small asteroids to Earth's orbit (Tate 2013). Eventually, this company plans to develop a tug to drag larger asteroids to the vicinity of the Earth (Tate 2013). Once there production with 3D printers would commence. As the second division would be undertaking similar activities, the Joint Stock Company may eventually acquire Deep Space Industries if this company did actually demonstrate a critical competence with these activities. The case would be similar for Planetary Resources.

As the asteroid mining industry is still nascent, there is very little legal basis for the right of property for these asteroids. While there is some vague language in previous UN accords about space being the preserve of all mankind and all nations having an equal stake in the resources in the heavens, the legalities surrounding the minerals located in these asteroids are still under developed. The resources to be found on these asteroids, such as titanium and iron, will most definitely be the building blocks of space civilization enabling the economic artifice to be developed without having to painstakingly bring materials to space from the Earth (Larson 2014). However, while there has been a lack of Congressional action concerning property rights on asteroids, or anywhere else in the solar system, such political dynamics will not delay the company's exploitation of the resources of space. In this regard, the company will adopt the same political mindset that political theorists such as John Locke espoused—that if one mixes his labor with the land the produce becomes his property. In any event, the security personnel of the company will be so formidable that it could be impossible for any other nation or economic entity to effect any action against the company—nobody will be able to stop the company from exploiting the resources of space.

Therefore, with or without any political action being taken by the United States Congress or the United Nations, the second division can proceed to make the most use of all the materials to be found in asteroids to support the burgeoning space economy. By making the maximal use of such resources in space, the company can greatly reduce the amount of material that it must transport into space from the Earth (Larson 2014). Here, "water will be particularly important. Beyond sustaining human life, water can shield people from harmful radiation and serve as fuel for spacecraft. It can be separated into its two components to generate energy or be heated with focused energy from the sun" (Larson 2014).

Thus, the company will not be taking on any risk by moving out with operations in space before there is a legal basis for them because, as previously stated, due to the company's overwhelming security and economic superiority there will be no nation or economic entity that could enforce an adverse political action against the company. Simply put, there will be nothing that can stop the company from exploiting the resources of space to the fullest which will create a politico-economic dynamic that drastically reduces any politico-economic risk. While "there is currently no legal guarantee that the Company will be able to profit from the resources they mine," there will nevertheless be nothing that can stop the company from exploiting these resources and building civilization in space (Larson 2014).

However, into this legal void, the United States Congress has in fact attempted some legislative action. While still under developed, "the Asteroids Act also protects a company's operations on an asteroid from 'harmful interference'" (Larson 2014). "Although the bill doesn't yet define 'harmful interference,' it will probably define it similarly to the way that custom and law does for fishing on the high seas. A fishing vessel is free to fish a lucrative school without another vessel navigating in the way to cause harm to the first vessel. These two elements—ownership and protection from interference—are essential for any enterprise to realistically pursue mining anywhere, including outer space" (Larson 2014). In any event, such language or intent will not matter for the company because it will have such a large security preponderance in space that no other entity or nation will be able to interfere anyways.

Regardless of any legislation or treaties that any politicians may produce, the second division could move forward with the exploitation of all the precious metals that are found in asteroids, including "platinum, rhodium, iridium, rhenium, osmium, ruthenium, palladium, germanium and gold" (Thomas 2013). Indeed, Planetary Resources believes that some objects that may only be just 500 m wide could possibly possess reserves of platinum group metals in greater amounts than can be found on the Earth. Iron ore may be found in space in even greater abundance. What is more, one Wall Street firm believes that there is an asteroid 200 km wide in the "belt between Mars and Jupiter" that could possess 17 quadrillion tons of nickel iron, which would be enough to meet humanity's needs for millions of years (Thomas 2013).

However, again, it would be technically and financially impractical to ever import such minerals to the Earth from space. Therefore, the second division would seek to exploit such mineral deposits in pursuit of the construction of the economic artifice of space civilization. In this regard, water, and the hydrogen and oxygen that can be derived from it, will be just as important as these metals (Thomas 2013). Thus, for colonies on Mars, or elsewhere into deeper space, plying the resources to be found in the cosmos makes tremendous sense because these resources will be en route to future human settlements. In order to have colonies in these locations, it will be necessary to use local resources. The water that could be had on asteroids en route to these settlements will be critical if humans are still using chemical propulsion at this time (Thomas 2013). Once again "the energy released when hydrogen and oxygen combine to make water can power rockets. The presence of both elements in compounds found on asteroids offers scope to set up space factories to make fuel for missions to Mars and beyond as well as offering 'pit stops' to extend the lives of satellites" (Thomas 2013). Indeed, propellants for space-based assets have become a multi-billion dollar industry that will only grow ever larger as humanity extends its presence across the solar system. Like the plans that Deep Space Industries has to create oases from asteroids where air and propellant can be had, the second division will also seek to develop these in situ resources for weigh stations throughout the solar system, especially the inner solar system where humans will reside in the immediate future.

The potential of asteroids is virtually unlimited. Some have even pegged the value of one asteroid in the main belt between Mars and Jupiter to be as much as $19 trillion (Godwin 2014). Evidently, it contains significant platinum group metals. In this regard, developers have considered rationing the minerals, similar to DeBeers business model with diamonds, while others continue to stress opportunities to utilize these minerals in the emerging space economy (Godwin 2014). Probably, both methods will come into play as space civilization develops. But another extremely attractive financial opportunity that all the resources of space pose is the creation of a new currency in outer space backed by all the deposits of the minerals that could be mined from asteroids (Godwin 2014). In fact, there could be multiple depositories of these precious minerals—in cislunar space with minerals from near-Earth asteroids and in the main belt between Mars and Jupiter with minerals found in the asteroids there. "Two cubic feet (0.05 cubic meters) of platinum weighs approximately one ton of mass and was worth about $55 million at market prices" in 2014 (Godwin 2014). The second division could then store up massive amounts of minerals at these depositories to back a space currency. The currency could even be printed on the Earth and could be used in a wide array of economic exchanges (Godwin 2014). Perhaps the currency would even become so strong that it would eventually replace the dollar as the reserve currency of the world. "A currency backed by limitless precious resources, with more practical utility than gold, may be preferable to currencies backed by the good faith of governments perhaps incapable of balancing their own books" (Godwin 2014).

Mars and Beyond

Moving further out into space, the second division would prize Mars as a potential new settlement. While the lunar colony would be a boon because of its ability to support the exploration of near-Earth asteroids, a colony on Mars would be tremendously beneficial not only for its resources but also for its ability to support the second division's missions into deeper space. Besides, Mars has historically been a very important destination for the American space community because it might have conditions that would enable humans to inhabit it, even if it would never be a "short sleeve" environment. In any event, critically, the second division would have the resources to create a human civilization there and fulfill the dreams of many space enthusiasts.

There has been much speculation about whether humans could ever exist on Mars. Such notions have been the staple of science fiction for decades. However, scientific evidence is beginning to emerge to support the idea that humans could exist on Mars. NASA has been sending probes to the Red planet to determine this very question. The $2.5 billion Curiosity rover is one such probe (Vergano 2012). It has been traversing the rugged Martian landscape looking for signs that the planet could support human life. According to NASA administrator Charles Bolden Curiosity is the largest mission to the Red planet and is a precursor for human missions to Mars (Vergano 2012).

Under the Obama administration, NASA is planning to conduct human missions to Mars as part of an international effort sometime in the 2030s (Vergano 2012). However, if indeed Mars can support human life, and many, including Space X founder Elon Musk, believe it can, the second division could be in position to conduct missions to Mars far earlier. As soon as transportation spacecraft is ready, which could be in the space of a decade, the second division would have the financial resources to begin conducting a large amount of sophisticated missions to Mars at some point in the 2020s with the goal of establishing a human civilization on the Red planet. Indeed, there are American astronauts, past and present, who have no doubts that it will be possible to establish a human presence on Mars (Vergano 2012).

In this regard, "plans call for pre-fabricated Martian habitats to be launched two years before astronauts themselves begin their journey. Once secured on the surface, astronauts would spend their days on Mars brewing methane fuel and oxygen from water and carbon deposits that evidence suggests remain locked under the Martian soil" (Vergano 2012). While NASA missions to Mars in the 1970s, such as the Viking lander missions, painted a bleak picture for the prospect of Mars supporting life, more recent NASA missions have determined that there are in fact caves, water, and minerals in significant enough amounts on the planet to support the beginnings of a human civilization there (Vergano 2012).

Radiation will no doubt be a problem for humans on Mars. However, "the water that crews will carry to Mars will serve as shielding against deadly solar particles and cosmic rays. And once on the Red planet, the roofs' of habitats could have some 2 m of Martian soil for protection" (Moskvitch 2013).

As previously mentioned, the founder of Space X for one has no doubt that it will be possible for Mars to sustain life. However, Musk is on record suggesting that it may even be beneficial to "bioengineer new organisms that are better suited to living on Mars" (Mack 2013). "Musk compared the biological adaptations that could be required for humans or other organisms to live on Mars to the process of breeding cows to conform to our needs as a society. The problem is that modern cattle are the product of hundreds of generations of tweaking to various breeds, but future space explorers won't have the luxury of being bred for the job of colonizing Mars through the centuries" (Mack 2013). Such realities, however, may not hold back humans on the Red planet due to the tremendous advances in sciences and biological knowledge that will without doubt accompany human missions there. In this regard, the development of synthetic organisms may even be necessary to expand the human presence in the solar system beyond Mars past the asteroid belt (Mack 2013).

In any event, the first step towards the colonization of Mars is to shorten the tremendous amount of time required to travel there. As such, it will be necessary for humans to exploit the brief six month period of time when "the orbits of the Earth and Mars sync up every two years" to increase the practicality of travel to the Red planet (Mack 2013).

Beyond Mars, there is another destination in our solar system that may eventually be suitable for the second division to conduct human missions to. In this regard, one of Jupiter's moons, Europa, "has intrigued scientists since the late 1970s, when the Voyager spacecraft first glimpsed its strangely fractured surface. The Galileo spacecraft, despite its crippled main antenna, confirmed the existence of water and an iceberg-like disjointed surface in the 1990s, but raised more questions than it answered: there is probably water in liquid form, but what is the exact extent of it? What is its chemistry? Is Europa habitable now or in the past? What could live there? Is anything living there" (Dreier 2014)? If, indeed, the answers to these questions can be positively confirmed the second division will no doubt eventually establish a settlement on Europa as well.

Chapter 4
The Third Division—Spillover

Perhaps the most beautiful aspect of space civilization would be the massive manufacturing base that would return to American soil. At this moment in history Americans have a competitive advantage over other nations in terms of the sophistication and quality of the space systems they create. Therefore, while the company would likely strategically allocate some of the trillions that were getting reinvested into space in other countries around the world, the American economy would obviously have the largest portion of the entire allocation—perhaps as much as a trillion dollars. This would be twice the size of the Pentagon's 2015 budget. Imagine the jobs this expenditure would create. It could create tens of millions of high tech manufacturing jobs. Not only would there be that many jobs created, but they would be of the best quality with the highest levels of pay.

This outlay would create an industrial renaissance in America, especially in areas such as the Midwestern Rust Belt and the Deep South. It would mobilize the American economy for space in a fashion strikingly similar to how it was mobilized for World War II. Assembly lines all around the country would be churning out systems for space. The most innovative and sophisticated methods of manufacturing would have to be developed to produce these space systems. These could then be applied across the rest of the economy in other manufacturing industries. In this way, the United States could regain its former industrial might.

Also of considerable beauty would be the technological explosion that would ensue in American society. All of the technology that the second division was developing to create space civilization would "spillover" across terrestrial society and have applications in the global economy. Thus, the third division would be a vehicle for bringing all the dual use, synergistic technologies for space to terrestrial markets. This would hugely increase revenue streams to the joint stock company.

As the second division develops the means to carry humans further and further into the solar system it would constantly be innovating the most sophisticated

T. Brown, *The Digital Galactic Complex*, SpringerBriefs in Political Science 30,
DOI 10.1007/978-3-319-22386-5_4

information technologies, infrastructural technologies, biomedical technologies, and a variety of other technologies, etc. In this regard, the drive to colonize space would create an infinite well of technological innovation that could forever spur the global economy to greater heights than would otherwise be the case without this effort in the cosmos. Not only would there be a wave of innovation way beyond electric vehicles caused by the necessity of the global economy to adapt to the new energy technology, there would be massive waves of scientific and technological breakthrough spilling over from space civilization. This would serve to create millions more high tech manufacturing jobs and would boost the American economy to the greatest of heights.

The Effect of Technology Development on the Broader Economy

Technology is a phenomenon that constitutes know-how, finished products, and everything that is derived from the technology in question. According to Hertzfeld (1978), "dictionaries and the aerospace industry define it effectively as know-how applied to practical purposes" (1). Technology is typically distinct from scientific knowledge, rather consisting in the application of such knowledge to systems and processes. According to certain reports from the formative years of a nascent space industry technology can be defined "as know-how for design, development, manufacturing, quality control and testing, performance analysis, maintenance and repairs, etc. Closely associated with such know-how may be instrumentation and basic knowledge necessary to its use" (Hertzfeld 1978, p. 1). However, in an increasingly globalized world that is being increasingly linked at an exponential rate by ever sophisticated communications and information processes basic knowledge is being disseminated in widespread fashion across the globe with ever fewer restrictions. Therefore, it is becoming increasingly difficult to control the diffusion of technology and such controls should only be attempted in cases where the fundamental technical understanding of it is vital to its application or if it remains unique in the circumstances in which it occurred and the knowledge of which is not yet widespread. Regardless, for our purposes, technology should be understood as know-how, combined with equipment and scientific understanding necessary for its operation, typically at sophisticated levels (Hertzfeld 1978).

To expand on this concept as it applies to the economy technology may also be understood as a "body of knowledge concerning how society's resources can be combined to yield economic goods and services" (Mathematica 1976, p. 1). NASA's technological developments over its history constitute NASA's contribution to American society's body of technical knowledge. This contribution was quite considerable, a fact which should give a good indication of how impressive will be the joint stock company's contribution to American society's technological gains in the twenty-first century.

Technology plays an indispensable role in the economic activity of nations. In this regard, "the degree to which a nation can satisfy its collective and individual wants is dependent upon the wealth of the nation and its citizens. The accumulation of economic wherewithal is obtained through combinations of labor, capital, and technology. All three inputs are essential but it is through technological progress that the productivity of labor and capital are increased to obtain more output per unit of input and, consequently, greater per capita wealth" (Midwest Research Institute 1971, p. 1). Thus, technological gains have an important impact on the economy, ultimately serving to increase the productive forces inherent in society which in turn increases accumulations of capital and the wealth of the nation.

Generally speaking, the economic process entails the transformation of a nation's supply of resources into products and then involves the distribution of these products in exchanges in the marketplace. Such functions create economic benefits by providing members of society with the ability to consume a wide variety of products that they desire. By ever increasing the sophistication of their technology, societies are able to multiply these benefits by deriving ever "more from the same stock of resources" (Mathematica 1976, p. 2). The ability to quantify economic benefits has been hotly contested in the field of economics for a long time. In the 1970s, shortly after the end of the Apollo program, the consensus was that "the most widely accepted principle for evaluating economic benefits is founded on individuals' 'willingness to pay' to move from a 'less' to a 'more' preferred state. What this principle translates into for the purposes here is that the benefits of technological change can be measured as the cost savings generated by new or improved production processes plus the extra value that consumers attach to new or improved final products. Therefore, by determining how cost and demand for various products are affected by specific technological advances, one can estimate the benefits of these advances" (Mathematica 1976, p. 3).

R&D

Furthermore, the R&D investments that the American space program has undertaken from its inception to the present have had a tremendous impact on the nation's economy and at the global level. Without doubt, the pace of national and global technological change has been greatly increased by the American space effort, which has served to generate tremendous economic growth in addition to an array of impressive new products and vast wealth. Entirely new industries have come into being as a result of the American space effort. New technologies have moved from initial development in laboratories to the marketplace far more rapidly than they would have in the absence of the space effort. This has resulted in an abundance of new jobs and income and has drastically "improved the quality of life" in American society (Hertzfeld 1998, p. 385). Clearly the impact of the joint stock company's massive outlays in space will substantially increase the amount of space R&D taking place in the United States and around the world and will in turn redouble the tremendous effects that such R&D has had on the advancement of technology and the sophistication of the global economy.

There can be no doubt that the kinds of R&D investments that have been made by the American space program have had a high correlation with the advancement of many innovations that have been found to be critical to the maintenance of the nation's future competiveness (The WEFA Group 1994). Some of these investments have led to technologies such as "artificial intelligence, advanced robotics, optical communication, and advanced computers. Because of their cutting-edge nature, NASA programs are highly conducive to promoting technology advancements. This store of technology is an important national resource because it can be adapted to develop new products and processes" (The WEFA Group 1994, p. 3). Indeed, many of the jobs of the future will arise out of entirely new industries that will proceed from developments in laboratories at the company. These laboratories will not only be found on Earth but also in orbit. As such, indefinite periods of microgravity and the other unique characteristics of orbit will without doubt enable the development and manufacture of highly sophisticated products, ranging from "medical preparations for fighting some of our most widespread diseases; alloys stronger yet lighter than any presently known; electronic components for faster and smarter computers and better electronic machines than are now available, and systems for almost universal information availability to increase the diffusion of knowledge" (Fuller 1984, p. 4). Already the tremendous view from orbit has enabled tremendous advances in "meteorology, cartography, reconnaissance, communications, earth sensing, and wide-area navigation—all of these have already brought important benefits to the people on Earth" (Rockwell 1978, p. 1).

After the Apollo program, and as a result of economic difficulties in the 1970s, NASA undertook to conduct various analyses on the economic impact of their R&D investments because with growing budget constraints it became increasingly necessary to justify their need for more and more resources. This reality was certainly not an issue before Apollo and the race for the Moon, but it definitely was by the end of Cold War. Regardless, the studies found that the total investment that the US government put into the civilian space effort through NASA alone amounted to over $200 billion over the course of almost 40 years. If translated to 1997 dollars that amount would be over $450 billion (Hertzfeld 1998). Such investments pump money into the economy and create jobs. The company's efforts will look quite significant in comparison as the company will be investing trillions into space on an annual basis, obviously to substantial technological effect.

Furthermore, in comparing civilian space expenditure to other forms of federal spending the type of occupations and the regions in question may differ, such as for welfare benefits, public housing, or new roads, but the direct result of these expenditures have similar multiplier effects while the NASA expenditure increased society's technology and in turn its productivity (Hertzfeld 1998). According to the Midwest Research Institute (1971), the addition of technological gains between 1949 and 1968 created 20 % more output than would otherwise have been created without NASA. These statistics bode extremely well for the R&D investments that the joint stock company would make in the future.

Indeed, according to various studies undertaken in the 1970s with the aim of supporting NASA's R&D investments there is significant evidence to support the view that developing the national technical aptitude necessary to maintain a thriving space program will create above average economic effects as a result of all the technological advances that will flow from the effort (Midwest Research Institute 1971). While it may remain difficult to produce an exact multiplier for industry-level technology that flows from space R&D, the technology that does in fact spread from space R&D is so significant that it is reasonable to maintain that sophisticated technological undertakings, such as space activities, ultimately have a massive impact on national productivity (Midwest Research Institute 1971). What differentiates R&D for space activities from other types of R&D is that R&D for space increases the productive capacity for the broader economy through the creation of entirely new and innovative technologies. Such new and innovative technologies necessarily serve to increase the efficiency of production methods that are already in existence, but at the same time also introduce new goods and services that open "new markets, industries, and opportunities" while drastically improving the quality of life for society (Hertzfeld 1998, 386).

To expand on the effects of R&D spending, "the demand effects are primarily short-run in nature and have consequences similar to that of other types of government spending. The supply effects, which represent the results of a higher rate of technological growth manifested through a larger total productive capacity, are long-run in nature and have consequences very dissimilar to that of general types of government spending" (Chase 1976, p. 1). According to a Chase Econometric Associates, Inc (1976) study "a shift toward higher NASA spending within the framework of a constant level of total Federal expenditures creates jobs without raising the rate of inflation, and hence is more stabilizing in a recovery period than general government spending" (5).

R&D expenditure for space stimulates the rate at which technology advances and decreases "the rate of inflation for two reasons" (Chase 1976, p. 17). First, in the near term, it redistributes demand toward sophisticated technical sectors of the economy, which in turn increases productivity in aggregate for the overall economy. Consequently, space-based R&D expenditures are typically more stabilizing in a recession than other government forms of economic stimulus. Second, over the long term, it speeds up the rate of technological advance which increases the productive capacity of the overall economy (Chase 1976). "This improves labor productivity further, which results in lower unit labor costs and hence lower prices. A slower rate of inflation leads in turn to a more rapid rise in real disposable income permitting consumers to purchase the additional goods and services being produced and generating greater employment" (Chase 1976, p. 17).

Societal Effects of R&D

R&D for space is also different from other forms of government expenditure because it necessarily requires personnel with greater income and more sophisticated education that expect greater socioeconomic conditions from their immediate surroundings. "This in turn affects the quality of education, for example, available to new residents of the community as well as to old ones, resulting in greater levels of achievement by all" (Hough 1968, p. 1).

The totality of all the benefits and impacts of space R&D on the economy do not occur over night. The process of turning a concept into a promising good or service may take several years or even decades. So while it is easier to measure the effects of direct federal expenditures on a short term basis, it is far more difficult to measure the overall benefits of R&D for space as they are not always fully felt for long periods of time (Hertzfeld 1998). Nevertheless, there can be no doubt that the totality of the impact of space R&D is indeed tremendous for society. In this regard, the social effects of the joint stock company will most definitely increase the sophistication of societies all around the world.

Space and Environmental Symbiosis

What is more, there are often areas of interest that are extremely important to the public where industry has no desire to do R&D. In the commercial sector for space this has historically been the case when it comes to conserving the electromagnetic spectrum and the use of GEO (Hertzfeld 1978). Likewise most of private industry is not interested in conserving the environment. Proposals have been floated by politicians to introduce legislation that would impose costs on polluting, such as a carbon tax. However, the joint stock company, in the process of colonizing space, would necessarily be conducting R&D which would be producing technological effects that would create the capabilities of preserving the environment that the rest of industry had no interest in developing without having to impose onerous financial consequences arising out of any legislation or executive orders.

Indeed, the space colonization effort would have many synergies with environmental technology. One example is water utilization capabilities in space. As water is much more scarce in space than on Earth it is necessary to develop technologies that will allow astronauts to utilize water to the maximum possible extent. These technical methods and mechanisms will enable entire colonies on the Moon and in orbit to make the most efficient use of water that can be found on the Moon or transported into space. Obviously, as the colonies arise these methods and mechanisms will grow in sophistication. Once in development, or fully developed, these methods and mechanisms could spillover to arid parts of the Earth and enable populations in these locations to make more efficient use of their water supply. This is but one of many ways the space program could help create a terrestrial society living in greater harmony with the environment of Earth.

Indeed, all of the technology developed for space could improve the way in which humans interact with the environment of Earth. The capability to exist on other planets would mean the capability to exist in greater harmony with the home planet. So while the second division was busy building civilization in space, the third division could be equally busy building a space age civilization on Earth. The third division could undertake to replace much of the infrastructure around the world with the sophisticated technologies derived from the infrastructure being developed for space. In tandem with this massive real estate development, perhaps the United States government could devote some of its massive budget surplus to public works projects as well. It could easily replace roads, bridges, etc. while the third division was replacing real estate and the electrical grid. Quite obviously any effort to replace the electrical grid would involve a massive upgrade to its security architecture. In any event, the combination of these projects would have an infrastructural impact greater than that in which Augustus rebuilt Rome with marble.

Revolutionary Technologies Flowing from R&D

Other studies have identified a number of contributions that the American space program has made to the general economy. "These include: developing new knowledge; developing new technology; demonstrating the application of new technology for the first time; augmenting existing technology; applying existing technology in a new context; stimulating industry to acquire or develop new technology; identifying problem areas requiring further research; and creating new markets" (Robbins et al. 1972, p. 3).

In this regard, some major technical developments have been brought to the marketplace with limited governmental support or intervention. These include "communications satellites, the miniaturization of electronic components, large management computer software applications, and advanced composite materials" (Hertzfeld 1998, p. 386). Traditionally, the private sector has managed around two-thirds of space R&D under contractual agreements with the American government. As such, when a technical advance is produced that will easily increase the efficiency at which proven and existing markets are served, the private sector has already been part of the development and is poised to commit the resources necessary to bring the technology full cycle and ultimately to international markets (Hertzfeld 1998). This reality will necessarily mean that the joint stock company, as the centralized financial and technical catalyst for technological development, will be even more poised to exploit the technological advances that flow from tremendous expenditure on space R&D than has the rest of the American private sector in the past.

There can be no doubt that future activities to colonize the inner solar system will create entirely new industries from all of the spillover from space operations. An abundance of new and improved consumer products will spread massively all over the globe. Already "requirements posed by NASA programs like Apollo, planetary exploration, and the Shuttle have produced miniaturized electronics, power systems and components, automatic checkout equipment, computers and software, high-volume data processing and communication, guidance and control systems, high-strength materials—the list is virtually endless" (Midwest Research Institute 1988, p. 1).

A fantastic example of spillover from NASA R&D spending is digital technology. Due to the complex requirements of deep space exploration NASA undertook to develop the use "of error-correcting codes and data compression in processing digital signals for modern-day digital communication and data storage" (Midwest Research Institute 1988, p. 3). NASA's active involvement and commitment of both human and financial resources in this promising area of technology enabled massive breakthroughs in coding theory. Before NASA's commitment in this technical area the practicality of coding for communications purposes was not very great. But in order to avoid "adding weight, power, and complexity to spacecraft" NASA undertook to develop this technology long before it held the promise of commercial success (Midwest Research Institute 1988, p. 4). This is but one example of a technological advance that was rapidly sped up by space R&D. A plethora of entirely new industries and products have been derived from this development. The gains to productivity for the overall economy by this technological advance have been more than considerable. Meanwhile, the benefits to consumers and society in general may go beyond any efforts at quantification.

Another incredible example of revolutionary technology that has spilled over from space R&D concerns integrated circuits. "Prior to 1960, conventional electronic circuitry was based on the assembly of individual encapsulated circuit components such as transistors, resistors, capacitors, and diodes" (Mathematica 1976, p. 6). Once the civilian space effort was in full swing NASA R&D efforts resulted in a number of advances regarding integrated circuit technology. Personnel at NASA, in collaboration with other organizations under contract, were able to "combine these circuit functions on an inseparable, continuous basis" thus yielding far more sophisticated circuitry (Mathematica 1976, p. 6). As such, there were a multitude of advantages to be had with the more sophisticated integrated circuitry, including a vast reduction in size, "lower power consumption, increased speed of operation, improved reliability, and reduced cost per electronic function" (Mathematica 1976, p. 6). Not only were these enhanced features critical for applications necessary for the exploration of space, but also they had a tremendous impact on the technological base of society. The development of integrated circuits revolutionized the way electronic products were designed and produced. This breakthrough allowed a quantum leap in the quality of consumer electronic products. The magnitude of this effort not only had a tremendous impact on the productive capacity of the entire economy, but also significantly improved the quality of life of society by raising the sophistication of its technology.

Cryogenic multilayer insulation materials were yet another breakthrough yielded up for industry by NASA R&D. In order for NASA to proceed successfully with its mandate to explore the cosmos it was necessary for the organization to "minimize the weight, volume, and evaporative losses of gas used in launch and flight propulsion systems, life support systems, and power generation on board spacecraft" (Mathematica 1976, p. 5). Central to this effort was the development of significantly enhanced insulation systems. As a result of such developments the cryogenics industry experienced an unprecedented period of growth and substantial increases in technical sophistication. In this regard, the main benefits to the cryogenics industry of NASA R&D efforts were the cost savings made possible by breakthroughs in multilayer insulation. Such breakthroughs substituted for "the next best insulation material" at the time, perlite, which would have otherwise been used to "transport liquid hydrogen, liquid helium, and liquid nitrogen" (Mathematica 1976, pp. 5–6). This development not only enabled cost savings in the transportation of such liquids, but also resulted in cost savings when it was used to produce these chemical gases. This development was applied to a variety of infrastructure and other equipment in this regard. The cost savings for the cryogenics industry were to be found in the reduced boil-off loss in the transportation process, and the lighter weight of the tanks being transported (Mathematica 1976).

Other revolutionary breakthroughs from space R&D came in the form of software. "NASTRAN (NASA Structural Analysis) was a general purpose finite element computer software package for static and dynamic analysis of the behavior of elastic structures" (Mathematica 1976, p. 8). This software development was pioneered by NASA but was modified soon after for various industries. In particular, mechanical and civil engineers found the software to be quite useful. These product engineers found applications for the software in the auto and aerospace fields as well as for constructing bridges and modeling power plants. NASTRAN was a product of research done within and outside of NASA. It was primarily developed at Goddard Space Flight Center between 1965 and 1970 (Mathematica 1976). The development was without doubt a significant improvement over other software programs available at the time, adding substantially to society's store of technology and vastly increasing the productivity of the economy.

In addition to all of the "new construction and building techniques," or new materials like anti-fog glasses, NASA contributed to an explosion in medical breakthroughs with such advances as "heart pacemakers and other medical instrumentation" (Hertzfeld 1998, p. 398). Indeed, the medical field is one of the most important areas of technology that NASA has had an impact on. In this regard, much of the information technology breakthroughs that NASA was responsible for were combined with medical techniques. One example involves "the virtual explosion in medical use of such computer technology as digital-imaging techniques and the like. The microminiaturization of electronic circuits as well as mechanical features, are especially adaptable to medical needs. The development of the Programmable Implantable Medication System (PIMS) and its substantial potential, along with digital imaging used for both brain and whole body scans and subsequent diagnostic procedures are only partial evidence of the explosive growth of this type of technology in the future" (Chapman et al 1989, pp. 26–27).

Digital technology, integrated circuits, and software represented a tremendous explosion for the American IT industry. Based on the historical fact of these developments it is safe to assume that yet another explosion in information technology will without doubt proceed from the joint stock company's efforts.

Speeding up Technological Advance

Concerning the effects of an increasing rate of technological advance on the productivity of the economy, studies undertaken in the 1970s also found that NASA's contributions served to spur technologies to develop before they otherwise would have if the efforts of industries and academics were not coordinated and accelerated by NASA's involvement (Hertzfeld 1998). Indeed, "the 'significance' of most of the NASA contributions was to have caused the technological advancement to occur at an earlier time than it would have occurred otherwise" (Robbins et al. 1972, p. 3). Indeed, there can be little doubt that space R&D drastically accelerates the pace of technological advance which creates significant effects for the overall economy that can be quantified in amounts substantially beyond the initial outlay on R&D that is the source of these effects. In fact, one such study performed by Mathematica found that the economic impact of NASA funding for R&D in the 1970s was ultimately more than NASA's annual budget (Hertzfeld 1998). The primary effect of the R&D was the speeding up of the development of technologies that otherwise would have taken far longer for industry or academia to create. Quite clearly, the increasingly centralized effort that the joint stock company would constitute, in combination with its increasingly massive resources, will only serve to spur the rate of technological advance far more greatly in the future than was the case with the American national space effort in the past.

The R&D Process

But while there can be no doubt about the effects of space R&D on the technological development of society, there is still a measure of difficulty in accurately attributing all the details of all the developments to particular individuals working at NASA over the course of the organizations history. "The crux of the matter is that few technological advances of any importance originate through the efforts of only a single person, group, or organization. Rather, numerous groups are likely to be at work on various aspects of the technology in a complex interacting way" (Robbins et al. 1972, p. 26). The difficulty in measuring technological developments in aggregate, or even with specific technologies, is due to the fact that innovative ideas are transferred in unpredictable ways, often without a way to trace how the process unfolds. Experiments take place and knowledge is then diffused along multiple paths making it nearly impossible to trace its path from the inception of an idea to the end product. Measuring an individual's contribution on a technology whose sales contribute to GDP then becomes virtually impossible (Hertzfeld 1998).

Expanding further, the research process complicates the picture even more. The research process can be characterized as a complex interaction of individuals and groups (Mathematica 1976). Often research is undertaken by teams. In attempting to solve a particular problem necessary for a technology to advance individuals usually build on the work previously done by others. In this fashion much research is really a combination or compilation. Therefore, much of the credit for the development of most technologies should be shared by multiple people or even multiple organizations (Mathematica 1976).

Perhaps this process can be best described by relating how a critical breakthrough in nuclear physics unfolded. At the end of 1938 Otto Han and Fritz Strassman were engaged with experiments that led to the discovery of nuclear fission. However, as it happened Enrico Fermi was performing similar experiments in Rome in 1934 while Irene Joliot-Curie and others were performing them in Paris in 1937 and 1938. Meanwhile at Berkeley Philip Abelson was also engaged with the same work (Robbins et al. 1972). "Knowledge flowed freely among the several research teams, influencing hypotheses and experimental design. In this multiple-path environment, it seems fair to say that if Hahn and Strassmann had not achieved their momentous insight when they did, someone else surely would have done so later, and in all probability not much later" (Robbins et al. 1972, p. 26). When considering the extent of NASA contributions on the technological base of American society it must be said that parallel and prior research was ongoing at the same time by other organizations in government, industry, and academia. However, the significance of NASA's contribution is that it necessarily served to increase the coordination of all of these efforts and in turn speeded up the pace of development. This dynamic will without doubt be multiplied by the increasingly centralized way in which the company will be catalyzing scientific breakthrough.

Perhaps the most critical aspect about the impact of the national space effort on spurring the development of technology is that it creates a demand for technologies that have yet to be developed or need more sophistication. The impact of NASA on the technological development of the United States was to spur technical cadres and organizations to strive for greater advances to fill the critical needs of the space program (Robbins et al. 1972). The developments that have heretofore taken place are typically the result of a combination of both NASA and industry efforts. These elements have worked in close combination, necessarily aiding advances as they unfold (Robbins et al. 1972). In this regard, the joint stock company will easily be able to coordinate and accelerate technological advance far more than has historically been the case.

Most of NASA's contributions to the United States' store of technology have been detailed by Spinoff magazine. However, Spinoff was never able to include all of the technical advances that NASA has been responsible for. This was so because of the aforementioned difficulty of tracing all of the technical progress back to its origins for the American public (Chapman 1989). This was the case with the manifold ways in which NASTRAN was put to use. At first, NASTRAN was utilized by NASA to structurally analyze the biggest rockets. But it was then modified thousands of times for various industrial applications (Chapman 1989).

Additionally, it was virtually impossible to account for all the technology that was derived from the national space effort and seamlessly embedded into a vast array of end products. "That is, the original NASA-furnished technology may have been the basis for a series of modifications during which the original technology, now embedded in the changes, has been 'lost' as to its origins" (Chapman et al. 1989, p. 1). In the mid-1970s, Congress directed NASA to also focus on ground propulsion as a result of the OPEC crisis. This dynamic would undoubtedly have been the case with any technical advances that would have proceeded from such an initiative.

NASA has also made substantial contributions to the American technological base beyond its R&D efforts by lending its unique capabilities at no cost to elements of American industry. One such example was in the early 1980s when McDonnell Douglas, an aerospace company, combined with Johnson and Johnson, a pharmaceutical company, to study the manufacture of new drugs in space. The initiative was intended to develop an electrophoresis instrument (Hertzfeld 1998). NASA agreed to lend the Shuttle to the effort at no cost so that the firms could test the equipment in space. Ultimately, the instrument was successful in separating chemicals in a near weightless environment that otherwise would not have easily been separated on the Earth (Hertzfeld 1998).

The Interrelationship Between the Space and Global Economies

As the world moves forward further into the twenty-first century, there are many challenges facing humanity. One such issue is uneven population growth that is expanding in the underdeveloped nations of the third world. As birth rates continue to drop in the industrialized West, they are growing substantially in developing nations. In this regard, the average age of the populations in the developing world is significantly lower than in the industrialized West. By way of comparison, the average age in the industrialized West is steadily increasing and significantly increased outlays for medical and old-age entitlements will be necessary to support such demographics. What is more, "the developed world has reached the time when each couple replaces itself with only two" children, or in some cases even less (Rockwell 1978, p. 1). Across the rest of the world, however, it is the opposite, as birth rates remain high and will likely remain so into the future. Such dynamics necessarily mean that most of the globe's population will remain in the developing world indefinitely. If the world continues on its present course, these dynamics will mean that the Earth will have to support a much higher population later in the twenty-first century. "Most experts are convinced that the best way to cut population growth rates is to develop a healthy worldwide economy. Emerging affluence has always been accompanied by reductions in population growth rates" (Rockwell 1978, p. 2).

One way to a robust global economy is through expanded trade. Indeed, the effects of globalization are serving to lift impoverished populations in Asia, Africa, and Latin America considerably. Their ability to export significantly to China, and China's ability to export significantly to the West, has been massively beneficial for the socioeconomic status of these developing populations. While the negative import–export balance has been deleterious for the United States, it has been a tremendous boon both directly and indirectly for all the economic links of the developing the world.

Regarding trade balances, if the total public and private investment in the United States and the rest of the West into space were to create a sustainable economy in the cosmos, this could create a new and emerging market for the West to export to. With Western exports to space surging, this would provide an ample opportunity for the developing world to increase their exports to the West and in turn spur further economic growth.

However, while Western R&D expenditures in space are creating a new market to export to, all of the products that would be derived from the spillover (as detailed here) could also find markets in the developing world, resulting in more evenly balanced trade all around. The increase in exports to the West would no doubt contribute to economic growth in these regions, thus increasing their purchasing power. This would enable these regions to import the advanced technology spilling over from space. The importation of such goods and services would no doubt raise their standards of living and vastly improve their quality of life. In this way, the third world could be lifted onto the socioeconomic status of the first world. These effects would necessarily serve to create advanced new jobs everywhere, including the mature economies of the developed West.

Increasingly robust global trade would easily secure the uninterrupted flow of raw materials for nations around the world. Evening trade balances between the developing and developed worlds would also enable increasing numbers of nations to sustain the trade in these materials. With adequate exports to pay for these materials all nations would then arrive at inclining economic positions.

Due to high labor rates, advanced technology products are essentially the only staples that Western countries can export competitively. Unfortunately, decades ago "other countries of the world started making unusually heavy investments in research so that many high-technology items that were once solid American exports became common imports. Television sets, steel ingots, precision optics, and automobiles are a few obvious examples. Thus, the only way we can maintain a positive balance of trade is to increase worker productivity or to stay in the forefront of advanced technology. Space industrialization offers many possibilities for exercising both of these important options" (Rockwell 1978, p. 4).

Indeed, without doubt both the first and third worlds will benefit massively from space industrialization. Regarding developing nations with emerging economies it is entirely possible to create credible scenarios which move these nations onto first world, industrialized status with all the trappings of Western standard of living amenities within a few short decades. What is more, space industrialization may allow this feat to be achieved without substantial local or global harm to the environment (Science Applications Inc. 1978). The joint stock company, as a centralized organization, can coordinate all of these efforts more closely and thus amplify the efforts and in turn speed up progress.

The Company's Social Altruism

Traditionally NASA has possessed unique technical and organizational capabilities that could have been necessary to meet the needs of pressing social problems, either directly or indirectly. "These were generally in the categories of specific technology, systems engineering, and organization and management" (Packer 1969, p. 2). In this regard, the joint stock company would now be in a position to meet many of the new challenges that society is presented with with just a small amount of its considerable resources. It could easily apply many of its dynamic technologies to problems besetting the public.

Historically, there has been a deep reservoir of altruism in NASA's organization that made members of that organization supportive of tackling these public challenges (Packer 1962). The same will most likely be true of the joint stock company. As such "an overt organizational step toward social application of technology, no matter how restricted and cautious, would most likely generate considerable enthusiasm" at the company (Packer 1969, p. 2). While some elements of the organization may oppose such steps due to resistance to change, administrative issues, and how frustrating the problems could be much of the organization would likely support any initiatives that involved application of the company's technology to intractable socioeconomic problems.

In the past elements of NASA found no problem with "extending NASA's technology charter to include direct participation in civil problems of national scope" (Packer 1969, p. 2). NASA was responsible for studying and exploiting perilous environments for humans including "zero gravity, underseas, radiation, aeronautical flight conditions, closed life cycle, extreme temperatures, vacuum etc" (Packer 1969, p. 2). In this regard, these elements thought it easy to transition much of the technical and organizational capabilities developed to surmount these difficult environments to intractable natural or social problems including "crowding, malnourishment, air and water pollution, noise pollution, violence and insecurity, fear, economic dislocation, resource depletion, earthquakes, and destructive storms etc." (Packer 1969, p. 2).

Concerning the joint stock company most of these issues will be naturally resolved directly or indirectly by the effects of the company's operations without having to extend the scope of the third division's or any other division's mission. However, it would nevertheless be quite easy to apply the technical and organizational advances of the company to intractable social problems if it became necessary.

It wouldn't matter that NASA's and the joint stock company's space efforts cannot really be compared to social advances "the reality of the situation" would be that the Joint Stock company, like NASA before it, would be "perceived by the public as a ray of hope" in a potential leadership position poised to offer assistance to tackle intractable social problems (Packer 1969, p. 5).

One way that NASA addressed these social problems was by creating joint initiatives with other government agencies along with special-purpose research institutes. Examples included "urban systems engineering, crime technology, building systems research, highway safety, synthetic food research, and air traffic control" (Packer 1969, p. 8). Perhaps, the joint stock company would have similar public outreach initiatives with the aim of more effectively combatting the manifold social problems besetting society.

In any event, the socioeconomic impact of the joint stock company would mean that the US economy would be beyond full employment. With the trillion dollar outlay in space, the waves of innovation rippling across society, and the massive works projects there likely would not be enough workers in all of the United States to fill all of the jobs necessary to maintain the economy. Therefore, it would be necessary for American innovators to redouble their efforts at automation if American citizens wanted to continue to shop at Walmart and have fast food. This would easily be the most prosperous period in American history, and even in world history. Not only would this economic transformation take place in the United States, it would happen all over the world. Opportunity would abound for billions across the Earth.

Chapter 5
The Fourth Division—Finance

The company's financial resources would be beyond massive. The profits from energy sales alone could exceed $5 trillion. When factoring in all of the profits generated by the emerging space economy, as well as technology sales in the terrestrial economy, the company's bottom line would explode. It is impossible to calculate how big would be the profits from economic activities in a future space civilization, or from all the activities conducted to create a space age terrestrial civilization. However, it is highly likely that as the decades pass and space civilization comes into being, the value of the entire organization could eventually exceed a quadrillion dollars. Needless to say, the company would need a vehicle to manage its tremendous financial resources. Therefore, the company would create a fourth division that would be constituted to invest its considerable resources in the capital markets and provide consumer financing for the sale of all the space technology spilling over across the globe.

The fourth division would begin its operations from a baseline of energy sales. As previously discussed, once the artificial moons were on orbit the company could achieve a global energy monopoly with about 75 % of a $7 trillion market. This would mean in excess of $5 trillion in profits. Around $2 trillion would go to space expenditure and slightly over $1 trillion would go to the United States treasury in taxes. This would leave roughly $2 trillion for the fourth division to invest in the capital markets around the globe on an annual basis. This would dwarf the sovereign wealth funds of Asia and the Middle East. The impact on the global economy would be tremendous to say the least.

The Capital Markets

Regarding the global economy, it is a very dynamic aspect of modern society. It consists of all the organizational mechanisms that go into producing products and services, the circulation of such products and services through a medium

T. Brown, *The Digital Galactic Complex*, SpringerBriefs in Political Science 30,
DOI 10.1007/978-3-319-22386-5_5

of exchange (currency), the market forces that determine the allocation of the resources used to develop these products and services, and all of their consumption (Dunn and Morris 1988). In the United States, the broader Western world, and now increasingly in more and more parts of the globe the market is what characterizes modern economies. In fully developed market economies most of the largest organizations have equity, or stock, that is tradable on exchanges by members of their society. According to studies of the stock market in the 1960s "among stockholders, the advantages of stock ownership were placed in the following order: long-term capital gains, good dividends, good to have when general prices are going up, and opportunity for a quick profit. They indicated strongly a belief that common stock provided good protection against inflation. Adults who did not own stock reversed the order of advantages and said they did not know whether common stock provided good protection against inflation" (Leffler 1963, p. 8).

For most stockholders, the main objective of owning stock is to derive gains from the appreciation in value of the shares they hold. Concerns about deriving income from their equity, assets are largely secondary. Such a dynamic may be considered speculation. That is the hope of acquiring stock at a low price and then selling high. As such, there is a certain measure of risk that is taken with such activities (Leffler 1963).

When those seeking to derive gains from an appreciation in the value of their stock plan, their investment strategies the most central concern they typically have is about the future earnings potential of the companies whose stock they seek to acquire. In this regard, sometimes young companies that will not have strong earnings potential for many years are attractive to investors because; however, long it may take, extraordinary earnings potential is there. Often times investors seek to invest in less risky blue chip stocks because the earnings are, in fact, already there. Indeed, a very significant emphasis goes into the processes for forecasting future earnings for corporations as well as for the whole of the stock market because earnings are such an important factor in the valuation of stock (Bolten 2000). There are many different variables that affect formulas for the prediction of future earnings that experts must examine. Once a certain degree of understanding is derived from these processes forecasts for the performance of stocks are then disseminated on a massive scale throughout the world by a wide variety of professional analysts making use of the most advanced methods of telecommunications. These forecasts are what determine the professional consensus of how individual stocks and the broader market will perform (Bolten 2000).

The performance of stocks varies to a great degree based on what part of the business cycle the economy is currently in. However, in the 1960s "many studies were done to indicate the long run appreciation of common stocks. The minimum annual appreciation found was about 3 %. Add this to the 5 % annual return and the total gain becomes 8 %. At this rate, money doubled in nine years, disregarding taxes and commissions" (Leffler 1963, p. 29).

A huge advantage of stocks as an investment is the tremendous liquidity of the markets. Such vast liquidity allows investors to buy and sell shares rapidly and in large quantities at prices that vary little from current market prices (Leffler 1963).

Such dynamics have given rise to investors who have developed extremely sophisticated computer software that executes trades in microseconds taking advantage of the spreads in buy and sell orders. Such trends are the culmination of trading practices dating back to the beginnings of the first bourses in Europe. As technology has developed over the decades and centuries such short-term investing strategies have become increasingly attractive to investors in possession of extraordinarily sophisticated technology.

But even as late as the 1960s, there were investors, or stock operators, who were only interested in short-term trading strategies seeking quick profits, while other investors bought and held for the long term. "In a study of public transactions in September 1961, the New York Stock Exchange found that 9.5 % of the public transactions were closed out within thirty days; 28.5 %, between 30 days and 6 months; and 62.0 %, sometime after 6 months had passed" (Leffler 1963, p. 9). Speculative activities are what would properly characterize transactions that were completed in under 6 months. Obviously the intention of such activities is quick gains from fluid price actions as opposed to any business fundamentals that corporations were exhibiting (Leffler 1963). As previously discussed such types of trading strategies have only grown more sophisticated with the advance of technology. Computers and software that can execute trades in microseconds have become the culmination of such strategies. In the twenty-first century, the market is becoming increasingly dominated by such technologically sophisticated market participants and the more traditional investor interested in stocks for the long term is gradually becoming less significant.

Nevertheless, for long-term investors, one of the main reasons for acquiring stocks is because they are a hedge against inflation. In this regard, real estate also provides similar benefits. "Bonds and preferred stocks are hedges against deflation, but common stocks by reason of a long-term upward trend are far superior in protection against inflation. Over a long period of American history, inflation has been a greater danger to the investor than deflation. If one wants to maintain stability of purchasing power, the common stock is more likely to be a safe investment than the bond" (Leffler 1963, p. 29).

However, "common stocks are only inflation hedges when the rate of increase in expected earnings induced by inflation exceeds the rate of increase in interest rates also induced by inflation" (Bolten 2000, p. 57). This dynamic is precisely one reason why the company's energy technology is so significant for the global economy. The ability of the energy infrastructure to collapse the oil markets will put downward pressure on both inflation and interest rates at the same time as profits grow throughout the rest of the economy. In modern history higher oil prices, especially in a period of oil shock, have typically raised the costs for businesses across the board which causes higher inflation which in turn also causes interest rates to soar. Meanwhile consumer demand drops putting pressure on earnings. Such dynamics have typically resulted in a depressed economy and have caused stock markets to drop. However, with the advent of the artificial moons, the dynamic will be completely changed. The artificial moons will deliver enough energy supply for the whole of the globe at very inexpensive and stable prices putting downward pressure on inflation, and thus interest rates, and then in turn causing stock markets to rise.

Factors in Valuing the Market

As has been discussed, the conventional theory behind the valuation of stocks is their future earnings potential. In this regard, stock prices change as a result of expected changes in corporate earnings (Leffler 1963). However, there are many other methods inherent in the determination of the value of stocks. Such methods are largely psychological in nature. Here "the basic factor in the movement of stock prices is the rise and fall of trader and investor confidence in the future of stock prices, earnings, and dividends" (Leffler 1963, p. 493). Such valuations are based on market psychology and not the business fundamentals behind the stocks.

Market psychology is a rather amorphous factor in the valuation of stock markets. In this respect, "economics is not an exact science like physics or mathematics but rather a social and human oriented science. This means that it is concerned with the behavior of collective individuals. Which is why at times it can be inconsistent and irrational. Just like a crowd of people at a football game or even worse, people caught in an hotel fire" (Dunn and Morris 1988, p. 12)!

Concerning the psychology of the markets, participants are constantly bombarded with an overwhelming barrage of information from a wide array of sources. As the steady stream of news can be bewildering at times investors may often find it difficult to make a truly informed decision when acquiring stocks (Leffler 1963). Nevertheless, there are, however, some basic indicators that are of tremendous aid to investors trying to decipher all of the news about the markets. As the stock market is largely composed of players who try to forecast the performance of specific corporations, as well as the broader economy, there are a number of economic signals that can do much in the way of revealing the reasons for why the market performs as it does. These include steel production, automobile production, department store sales, import/export figures, and inventories etc. (Leffler 1963).

In addition to these indicators political news is also an increasingly important determinant in market activity. Any market participant who does not take account of the political news affecting both the domestic and international situations will not have a full assessment of all the factors impacting the markets. Indeed, more so than ever stock markets are increasingly linked to political developments around the world. They have become just as fundamental as the regular business aspects of the economy (Leffler 1963). Both local and national governments all around the globe are central to economic activity in so far as they engage in taxation, determine expenditures, and exercise significant economic control. As the populations of nations have increasingly looked to their governments to become more and more responsible for the performance of their economies political developments are gradually becoming some of the most important factors affecting national economies. Therefore, the policies of administrations, legislation from Congress, and decisions emanating from federal courts must all be accounted for in any prudent investment strategy (Leffler 1963).

During the Cold War, Western investors watched the Soviet Union warily out of financial necessity. In the twenty-first century it is terrorism, a rising China, and even a resurgent Russia that investors must be keenly aware of especially in an increasingly globalized marketplace. Any developments in this regard could very easily overshadow the financial activities of the day (Leffler 1963).

With the multitude of chaotic political developments that could easily impact the market place, in combination with any heightened sense of fear that an adverse signal from an indicator might induce, it is easy to see how investors may get spooked for any number of reasons and overreact to bad news or signals when making investment decisions. In any event, "there are hundreds of reasons why at any given time, the behavior of individuals in a society does not correspond with logic. But it must be stressed that this does not exist over a long period of time. Sooner or later global logic prevails over all (or almost all)" (Dunn and Morris 1988, p. 13).

Regardless of how long it takes for logic to prevail throughout the global markets "the delays generated by these economic deviations create the effect of a pendulum in our economy. In any one given year consumers can suddenly become very optimistic. They feel richer and they buy no matter what the price. The following year they become nervous and they increase their savings and buy much less" (Dunn and Morris 1988, p. 13).

Furthermore, the extent of the change in stock prices typically serves to induce even more fear in investors. Even prices that soar more than is reasonable for a particular stock may cause some investors to become anxious, but this is especially the case when stock prices plummet. When these dynamics occur investors will typically raise or lower their expectations on the yields of assets that are more risky (Marathe 1998). These processes can cause huge fluctuations in the prices of equity assets. When these processes take place psychological factors will usually play an increased role over and above that of fundamentals in the valuation of equity assets. Scholars have found little evidence of any way to take advantage of the psychological factors and get the better of the market with any consistency (Marathe 1998).

In fact, scholars believe these psychological dynamics to be in play in the case of the October 1987 crash. They believe the precipitous drop in stock prices in this market event was due to a speculative bubble that did not comport with fundamentals or any rational notions of market efficiency (Marathe 1998). Evidently in this market event investors sold stocks only because of the psychological fear that other investors would also sell stocks. Thus, investor psychology was the main factor behind the crash (Marathe 1998).

What is more, scholars have argued further that volatility in the markets is not always merely the result of logical thoughts pertaining to how investors expect dividends and interest rates to behave in the future. Thus, excess volatility is seldom the result of fundamentals in stock prices (Marathe 1998). In this regard, "the market is not an automaton. Rather it is a composite of all the hopes, fears, generalizations, forecasts, guesses, and analyses of the thousands upon thousands of individuals, firms, and corporations that deal in the market each day" (Leffler 1963, p. 64). Additionally, there is even more empirical evidence "that market volatility increases during recessionary time periods" (Marathe 1998, p. 40).

The psychological factors in play that increase volatility in recessionary time periods may necessitate that the fourth division becomes more involved in the global equity markets during these time periods and under these conditions. Without doubt the resources that the fourth division would be in possession of would make it possible for the company to have a tremendous impact in this regard.

Indeed, a number of scholars have noted the deleterious impact of excess volatility in the markets. Some have suggested that in cases where the market falls by at least 20 %, it may indicate that the market is not in fact performing well in its societal role as the ultimate allocator of the economy's resources. When such adverse market events occur tremendous doubt is then cast on the quality of the signals that the market is giving savers and businesses regarding the outlets they should seek for the investment of their capital (Marathe 1998).

During periods of irrational excess volatility due to exaggerated psychological factors, the market sends unfortunate negative signals to investors that would heighten "the risks of corporate investment. That is, if there is significantly more volatility in corporate stock prices than in the fundamental corporate earnings stream, stock price volatility could increase risks to market investors out of all proportion to the economic risks of corporate investment. This could raise the cost of capital to the firm and cause the equity prices to be lower than they should be and hence reduce corporate real investment" (Marathe 1998, p. 8). In such events it may become necessary for the fourth division to deploy its tremendous cash reserves across the global equity markets in such a way so as to reduce the effects of irrational excess volatility.

Indeed, for investment managers mainly concerned with the fundamentals of stocks, irrational excess volatility only increases the difficulty of executing orders. What is more, for executives of companies with publicly traded stock the deleterious effects of irrational excess volatility only increases the difficulty of executing their corporate operations. Exceptionally high volatility only puts unnecessary pressure on financial managers when having to explain all of the turbulence in the markets to their clients (Marathe 1998). Such dynamics unnecessarily cause society to lose confidence in the markets, and, indeed, even in that of capitalism. If investors know that it is possible for their hard earned money to evaporate quickly in losses of over 20 % in a single day, as was the case in October 1987, their investing decisions will be informed negatively by such possibilities with profound and unfortunate consequences for the capital markets (Marathe 1998). To avert such negative dynamics the company could act as the ultimate market maker with its massive cash reserves and stabilize the markets by massively buying when the market is massively selling, and then gradually selling as the market resumes its buy orders.

In addition, the company could have a tremendous impact on global interest rates as it will be printing its own currency backed by precious metals in space. In this regard, future expectations about earnings and risk in the global equity markets that impact stock fluctuations are determined by interest rates in a very significant way (Bolten 2000). Toward the end of the business cycle expected earnings tend to decrease at the same time that interest rates typically increase. The combination of these two factors tends to cause significant falls in the market (Bolten 2000). The company of course will once again have a stabilizing impact on this process; however, while the company is increasing interest rates, it could steadily and simultaneously increase its outlays in space thereby relieving pressure on corporate earnings across the world at the end of the business cycle which could ultimately lead to a very soft landing, perhaps even avoiding recessions altogether, before engaging in even larger outlays and more stabilizing activities in the capital markets.

The Company's Strategy on the Worldwide Equity Markets

Beyond the fourth division's impact on interest rates and its stabilizing fiscal activities, it would also execute a strategy in the global equity markets, as previously alluded to. In this regard, if the company were to hoard cash then 10 years after the artificial moons were constituted on orbit, it would have a cash position of $20 trillion. This 10-year period would coincide with the business cycle across the Earth. Thus, as the company's stockpile of cash was reaching and exceeding $20 trillion, the global economy could be entering into another cyclical recession. In such an event stock markets around the world would plummet. As such, the fourth division would be poised to make major financial moves and seize tremendous assets at bargain prices. In fact, the fourth division's resources would be so great that these moves could prevent stock exchanges from collapsing. The immensity of the company's cash position could be enough to create a floor under all of the stock exchanges around the globe. This would save the savings of many small investors with 401k's who would typically otherwise suffer devastating losses.

But before the company initiates its massive market activities it should first seek to gain control of the global financial system. In this regard, the global financial system, beyond central banks, is largely comprised of financial institutions. "These institutions organize and assure the movement of necessary capital for the production and exchange of goods and services among all players in the economy" (Dunn and Morris 1988, pp. 27–28). The combination of all of these institutions and their operations in all of their totality constitute most of the global equity markets. Indeed, in the 1960s "about half the stock of publicly owned corporations was held by institutional investors. Some of this, to be sure, was held by trustees for funds established by individuals or by firms acting as custodians for securities owned by their clients. Other institutions—such as insurance companies, pension funds, mutual savings banks, investment companies, and common trust funds—that owned stock provided indirect ownership to individual policy holders, account holders, shareholders, or trustors" (Leffler 1963, p. 9).

Therefore, if the company could take significant stakes in most of the largest and mid-sized financial institutions around the world it would be in a position to gain a large measure of control over the global equity markets. To gain such control, the company would merely have to invest just a few trillion dollars in all of these institutions around the world, a completely manageable prospect for a company generating as much as $5 trillion in profits on an annual basis. Once such control was obtained the company would be poised to execute its market strategy. Once the takeover of institutional investors—investment banks, insurance companies etc.—was complete, the company could marshal their investment resources on the stock exchanges to be deployed in combination with the company's massive reserves of cash.

Thus, during the next recession, after the company had established the artificial moons on orbit, when stock prices become increasingly risky, the fourth division would be poised to take a certain amount of risk out of the equation by constituting significant demand for shares on the exchanges by buying stocks on a massive scale with its cash reserves. In this regard, the fourth division would be more than the ultimate market maker. In addition, investors could continue to anticipate all of the future earnings potential of publicly traded companies around the world after the recession is over. This dynamic will enable investors to continue to have confidence in the stock market during a recession.

Without doubt the fourth division would be manipulating the stock exchanges across the globe, but only in a positive way—the company would prevent their collapse. "Manipulation is an artificial control of security prices; it is an attempt to force securities to sell at prices either above or below those that would exist as a result of the normal operations of supply and demand" (Leffler 1963, p. 446). While most manipulators in the past sought "to profit by creating fictitious prices at the expense of the general trading public" the company would be seeking to stabilize the global equity markets in periods of irrational excess volatility induced by recessions (Leffler 1963, p. 446).

As such, while there are various objectives of manipulation, some of them nefarious, the company's manipulative activities would be largely altruistic. The main objective of the company, to increase the price of stock offered to investors when stock markets were crashing, would only be undertaken by the company to prevent the wealth of other investors from evaporating in periods of excess volatility. These activities would necessarily serve to stabilize the global equity markets. The main effect would be to ensure that small and unsophisticated investors could receive a better price for their stocks in trading operations than if the stocks were allowed to settle at lower levels induced by excess volatility (Leffler 1963). In fact, a similar process occurs when investment banks underwrite new securities. "It is an accepted practice which is permitted by the SEC provided the public is fully informed that such operations are contemplated" (Leffler 1963, p. 446). Indeed, the main objective of the fourth division would be to counteract the manipulative activities of other large and sophisticated traders who were seeking to depress stocks by short selling in bear markets and making massive profits at the expense of unsophisticated small investors who saw their wealth evaporate in recessions.

To this effect, with $20 trillion in cash reserves, the ability to print currency, and control over all the large and mid-sized financial institutions across the globe the company would be in position to "corner" all of the stock exchanges around the world in the next recession.

> A corner is a speculative situation in which the ownership of outstanding shares becomes so concentrated that short sellers are unable to secure stock except from the owner group. It grows out of a price rise in a stock, whether natural or manipulated. As the stock rises in value, it is short sold by speculators who feel that it is too high and is certain to decline. They borrow the stock to short sell; it is often loaned by the controlling group. Eventually they attempt to cover by buying back the stock which they have sold short. Since the controlling group is the only one that has any stock to sell, the short sellers are forced to settle with them at whatever price the group dictates. At this point the short sellers are 'cornered'...In these instances the manipulators freely loaned stock to the short sellers. As the price rose, short sellers continued to sell while the manipulators purchased all stock offered on the market. The stock was reloaned for more short selling. Eventually more stock would often be on loan than was actually outstanding. When this situation had continued to a point where the short sellers sought to cover either as a precautionary measure or in desperation, the manipulators called the stock loans. The corner was complete and the short sellers had no recourse but to settle (Leffler 1963, pp. 447–448).

$20 trillion in cash reserves and control over the global financial system should be enough to affect the global "corner," but if necessary the company could always print more money to complete the action. In any event, these activities would prevent large and sophisticated investors from shorting the markets and causing excess volatility that would lead to substantial losses for small and unsophisticated investors during recessions. In short, the company's financial power and massive cash reserves will be able to prevent all of the stock exchanges around the world from collapsing in the next recession.

Obviously, after this process took place the company would have a tremendous stake in all of the stock exchanges around the world. This position would allow the company to dominate the global economy. The political ramifications of this position would be tremendous as well. This position would allow the company to shape the politics of all nations.

The Fourth Division's Central Banking Function

In addition to such powerful market activities, the company will be in position to alleviate economic pressure in the next recession with moves to control the globe's money supply. In this regard, while the company would never need to raise money by offering bonds, perhaps it would create bonds as a safe haven investment for the global financial markets. Thus, by offering bonds, as well as a currency backed by precious metals in space, the company would be in position to have a tremendous impact on the global economy in recessions.

Currently, the "long interest rate on the long-term US Treasury Bond" is "the cornerstone for all investment alternatives" in a recessionary period (Bolten 2000, p. 12). There is virtually no risk of default with these securities. But once the joint-stock company is established, has commenced the energy operation, and has mined enough precious metals in space to store in vaults in the medium to back a currency, it will then be in position to offer bonds in a similar fashion as the US government. Then its "bondholders will always be assured of getting their interest and redemption payments" as is the case with US Treasury Bonds (Bolten 2000, p. 13). While "the US Government cannot run out of money because it prints more money in never-ending amounts" the company will be in position to do the same (Bolten 2000, p. 13).

Indeed, very easily the company's currency could displace the dollar as the reserve currency of the Earth. The fourth division would then become the principal financial institution of not only space civilization, but also perhaps the rest of the globe as well, in much the same way that the Federal Reserve Bank is the principle financial institution of the United States (Dunn and Morris 1988). In this capacity, the fourth division would not only be the personal bank of the company but it would also be the central bank of the world. Therefore, the fourth division would control the world's credit and money supply. It would issue money for the world, as well as civilization in space and would manage the finances and banking activities of the company. It would "serve as the official bank to all other banks, have the role of administering the globe's debt, and control the globe's credit level" (Dunn and Morris 1988, p. 28). The fourth division would control the credit of the globe by purchasing and selling the company's bonds in the same manner that the Fed purchases and sells the US government's treasury instruments (Dunn and Morris 1988). In this capacity, the fourth division would be in position to largely determine the rates of interest around the world. This would be a very important role for the fourth division because it would allow the company "to influence the movement of the" global economy, especially in recessionary periods (Dunn and Morris 1988, p. 28). Thus, the function of the fourth division on the world stage would be much like that of the Fed in the United States in so far as it would "foster a flow of credit and money that will facilitate orderly economic growth, a stable currency, and long run balance in international payments" (Board of Governors 1963, p. 1).

For the Fed "the principle means by which monetary policy achieves its objectives are open market purchases and sales, mainly of US government securities but also of foreign currencies; discount operations; and changes in reserve requirements" (Board of Governors 1963, p. 29). This will be the case for the fourth division as well. But only the fourth division will be dealing in the company's debt instruments. As such the company will need to establish a system of dealers in space and on the Earth who transact in the company's debt instruments as is the case with dealers of US government securities in the United States. In this regard, this system will allow the fourth division to both buy and sell the company's debt instruments on a massive scale and in turn have its intended impact on the globe's money supply.

The corresponding system presently in place in the United States operates in the following way: "when the Federal Open Market Account buys securities in the open market, the dealer receives a check on a Reserve Bank in payment, which he deposits in his own bank account. When a member bank receives these funds, they are credited to its reserve account with a Reserve Bank. Federal Reserve purchases of securities thus add to the reserve balances of member banks. Conversely, sales of securities reduce the reserve balances of member of banks. The resulting changes in reserve positions affect the ability of these banks to make loans and to buy securities" (Board of Governors 1963, p. 31). The system that the company will establish with its debt instruments will operate in a similar fashion.

With this system in place the fourth division will be positioned to determine both short-term and long-term interest rates. In this regard, the fourth division could increase the amount of its short-term debt instruments with interest rates set at an artificially high level by arbitrary processes to the general public which would have the intended effect of raising short-term interest rates. Simultaneously the fourth division could decrease the amount of long-term debt instruments with interest rates set at an artificially low level by arbitrary processes to the general public which would have the intended effect of lowering long-term interest rates (Board of Governors 1963). This dynamic could very easily be established if the fourth division determined that the global economy simultaneously required high short-term interest rates and low long-term interest rates. And if, of course, the opposite dynamic was required for the health of the global economy the fourth division could reverse the process and lower short-term interest rates while raising long-term interest rates in a similar fashion.

In addition, the fourth division would also seek to assure high levels of employment across the world through its financial activities. It would seek to accomplish this goal not only through its massive outlays in space on all soils of the globe but also through its control of the Earth's money supply. Indeed, as previously discussed the fourth division would be functioning for the world in a similar capacity as the Fed does for the United States. In this regard, as "bank reserves constitute both the legally required basis and the functional basis of bank deposits, changes in the reserve position of banks, therefore, directly affect the flow of bank credit and money" (Board of Governors 1963, p. 63). Therefore, in much the same fashion as the Fed makes use of its capacity to raise or lower "the volume and cost of reserves" in banks in the United States the fourth division would also leverage the status that its currency would have as the world's new reserve currency, as well as the status of its debt instruments, to control the money supply on a global basis and in turn lower and raise the volume of bank reserves in financial institutions around the world (Board of Governors 1963, p. 63). As the company would also have significant equity stakes in these financial institutions, this process would be undertaken with the utmost of ease in the smoothest fashion.

To the same extent that the Fed manages the money supply by issuing currency and servicing the US government's debt activities in an effort to attain its objectives of maximum employment, low inflation, and economic growth, the fourth division would, with its currency and debt activities, also adjust the globe's money supply to insure maximum global employment, low global inflation, and the highest possible levels of economic growth (Bolten 2000). In the process, the fourth division would create a long term and stable upward trend across all the global equity markets of the world. The fourth division of course would be helped significantly in this effort by the massive outlays that the second division would be having in space as well as the technological explosion that would be occurring around the world as a result of the third division's activities.

Nevertheless, the global economy would still be subject to similar economic cycles as in the past, although there would be a paradigm shift in economic activity. In this regard, lower levels of economic growth would be a relative phenomenon. There would still be more economic growth than there was at the bottom of past cycles, but cycles would be similar in so far as on a relative basis, there was lower economic growth at the end of the cycle than at the height of it. Therefore, during these periods of slower economic growth, it would still be necessary for the fourth division to increase the globe's money supply to boost economic activity and to stimulate global equity markets (Bolten 2000). In a similar but opposite fashion, during periods of tremendous bursts of economic growth the fourth division would be required to decrease the globe's money supply to cool off the global economy and worldwide equity markets (Bolten 2000). Such moves would necessarily impact global interest rates, raising or lowering them as required to increase or decrease credit as well as to stimulate or contract the worldwide equity markets. Obviously moves to decrease the money supply would be intended to tame inflation as well (Bolten 2000). Additionally, while the fourth division was making these movements with the globe's money supply, it would also be engaged in buying stocks on a massive scale during the market downturns, as previously discussed, while gradually selling off these positions as worldwide equity markets rose. Such activities would be intended to have a stabilizing effect on the worldwide equity markets.

What is more, through control of the global money supply, as well as through control of all the important financial institutions of the globe through its equity stakes, the company would be in an even more enviable position than that of the Fed to determine the levels of credit that financial institutions provide to society with loans and investments. Similar to the Fed, the company would also be able to determine the extent at which bank reserves were available as well as their cost. In this regard, "the reserve position of banks" across the Earth would be a significant factor in their ability "to extend credit and the rates of interest that borrowers" will have to pay to receive it (Board of Governors 1963, p. 9). The fourth division could control this dynamic indirectly by controlling the money supply and thus bank reserves, but also directly through the shares the company would hold in these institutions.

Indeed, "as banks extend credit by exchanging bank deposits for the various assets they acquire—promissory notes of businesses and consumers, mortgages on real estate, and government and other securities—they create demand deposits. These deposits can in turn be used to make payments by check or can be withdrawn for use in the form of currency. Ultimately, they may find their way into banks' time and savings deposits if the public prefers to save in that form. In any event, by affecting the volume of reserves available to member banks, the fourth division will have the power to influence the globe's overall credit situation and its money supply" (Board of Governors 1963, p. 9).

Furthermore, "the nominal interest rate component strictly related to compensation for the use of money, without regard to the risk of inflation, is the real interest rate. That rate is determined by the supply and demand for money" (Bolten 2000, p. 17). The company, as previously discussed, will be the most significant factor in the supply of money on a global basis, in fact dominating all aspects of it. While financial institutions around the world, through which the fourth division will operate, must make loans available for the global economy to function, at present they are not required to. Therefore, the process that the Fed currently controls in the United States can and does break down if in fact creditors do not provide the money that is given to them by the Fed to businesses and consumers in the form of loans (Bolten 2000). However, as the company will have stakes in these institutions it will be entirely possible for the company to compel them to make loans with the money that the fourth division provides.

While individuals around the world will still be required to save to augment the supply the company is creating for the market place, it will be less necessary for nations to have as much foreign direct investment in their markets as is presently required for growth because the company will be investing its considerable reserves from space in all nations around the globe. Thus, demands for money in economies around the world will easily be met from space even if their domestic savers are still not supplying enough (Bolten 2000).

Further along these lines, the fourth division of course will be by far the largest supplier of money and will, in the process, manage short-term interest rates around the world. In this regard, all of its moves will be closely scrutinized. Any change it makes in "the supply and/or demand for money" will cause corresponding fluctuations "in the real interest rate" (Bolten 2000, p. 17). These changes to interest rates will in turn affect stock markets, causing them to rise or fall in conjunction with the company's policies to stabilize the markets during periods of irrational excess volatility.

"The demand for money mostly reflects economic activity and the government budget surplus or deficit. As economic activity increases, the demand for money increases. Business and consumer demand rises for the purchase of homes, durable and non-durable goods, and for new plant and equipment. This does not occur uniformly throughout the economic/stock price cycle and has implications for portfolio sector rotation and individual common stock valuation" (Bolten 2000, p. 17). In the past, the US government has been a very significant factor in the demand for money. It has always gotten as much money as it has ever wanted irrespective of the interest rates it has had to pay. Whenever the US Treasury has needed more money, it has always borrowed. The effect has been to increase the yields of its debt instruments which in turn have led to increases in other rates that have historically put downward pressure on American stock markets (Bolten 2000). The company, however, will be unlikely to ever be under the same fiscal stress that the US government has been under in certain periods and will therefore be unlikely to ever require any loans. Thus, the company would necessarily raise interest rates on its debt instruments to artificially high levels if it ever needed to increase interest rates around the world in order to cool off the global economy. Conversely, the company would lower interest rates on its debt instruments to artificially low levels in order to lower interest rates around the world and stimulate the global economy. In this regard, interest rates on its debt instruments would be unlikely to ever be set by market forces related to any need for loans but rather by arbitrary mechanisms determined by officials in the fourth division in accordance with the needs of the global economy.

The fourth division will have many tools at its disposal to stabilize the global economy. As the mass of money grows around the world during low points in the business cycle credit will become easier, "encouraging purchases and investments. And of course the reverse" will be the case at the height of the business cycle when the company seeks to constrict the mass of money around the world (Dunn and Morris 1988, p. 28).

Additionally, much like the Fed, the fourth division will have a discount rate that it will apply to financial institutions around the world that deal directly with consumers. This will largely determine the rates of interest that these institutions will then apply to their customers (Dunn and Morris 1988). When the global economy heats up the fourth division will raise the discount rate which will in turn make loans more costly and begin to decelerate the amount of expenditures that actors are making throughout the economy (Dunn and Morris 1988). And of course when the fourth division wants to stimulate the global economy when it begins to ebb it will reduce the discount rate to encourage more expenditures. By reducing the discount rate, the fourth division will cause a corresponding reduction on "the interest paid on bank savings rates" as well as credit rates which will serve to encourage consumption throughout the global economy (Dunn and Morris 1988, p. 29).

The "reserve banking policy of the Company would attempt to provide a financial climate conducive to sustainable growth in output, employment, and consumption under conditions of relative stability in the average level of prices. However, these objectives cannot be attained through reliance on monetary policy alone. Their accomplishment also depends on fiscal and other governmental policies and on policies of private institutions and organizations" (Board of Governors 1963, pp. 127–128). Therefore, in a time of economic stress, the company's monetary policy would also be accompanied by an increase in outlays for the space economy with the company's hoard of cash reserves.

"To help avoid the dangers of economic downturn, reserve banking works to prevent speculative or otherwise unsustainable expansion of bank credit" (Board of Governors 1963, p. 128). "As market interest rates rise—a development that is reinforced by bank sales of securities—security prices decline and sales of securities may involve book losses. Banks are influenced to some extent by potential capital losses on the securities in their portfolios, and they hesitate to sell securities at a loss" (Board of Governors 1963, p. 131). The massive outlays of course would impact the business cycle tremendously allowing for continuous, uninterrupted growth. The constant innovation necessary to build space civilization will mean continuous growth.

Another activity of the fourth division would be floating the nonessential spillover technologies of the company onto the capital markets around the world. In this regard, the amount of technology that the second division would be creating, which the third division would be transitioning into terrestrial society, could make the third division's operations unwieldy. Therefore, in order to streamline the company as much as possible, a difficult proposition for such a mammoth company, the third division would create independent business units of its nonessential terrestrial technologies which the fourth division would spinoff onto the capital markets. Obviously the company would retain a stake in these independent businesses, but these technologies would then become available to independent investors all over the Earth. Indeed, the technological explosion at the company would be so great that it would become something of an assembly line for initial public offerings onto the capital markets. This dynamic would cause stock exchanges all around the world to explode. It would no doubt become a gilded era for financiers across the Earth.

Chapter 6
Grand Strategy

The capabilities of the joint-stock company would be more than considerable. As such the joint-stock company should make all due haste to execute a grand strategy to take over the Earth. To this end, the company should make the most use of the symbolic effects that the Solar Moons, all of its space activities, and its technological explosion have to offer. In this regard, the company should execute a psy-ops campaign with the Solar Moons and all of its technological dynamism. While this campaign is ongoing the debris service would continue to exercise control of space by offering protection to friendly nations and fowling the orbits of unfriendly nations.

Europe

The first objective in the European theater is of course to rush massive amounts of clean energy to European markets. The energy from the solar power satellites would undercut all competition, including Russian natural gas. What is more, as the energy would be clean solar it would be more politically attractive to the Europeans. Indeed, this exchange is certain to compromise the oil and natural gas revenues of Russia severely. In so doing, the solar power satellites would increase the company's leverage over Western Europe and alleviate the pressure that Russia is able to apply to the company's allies in Eastern Europe.

From the outset, the company should seek to deal heavy ideological blows to the Russian autocracy and oligarchy with a carefully crafted psy-ops campaign, as the Solar Moons and space civilization would be enormous concrete and visual proof of the power of democratic societies. Even if the Russian government could sensor state controlled media and erect great firewalls over the internet, the Russian government could never distort the brilliance of the Solar Moons and

T. Brown, *The Digital Galactic Complex*, SpringerBriefs in Political Science 30,
DOI 10.1007/978-3-319-22386-5_6

space civilization as it all shone forth from above, nor could it distort the view that captivated Russians had of the sky. It would be further damning evidence that while Russians remained unfree and continued to toil under heavy handed autocrats and oligarchs, free societies ultimately produced the creative ideas that spurred large and lasting economic growth, as well as massive comprehensive national power.

In addition to the psy-ops campaign, the precipitous drop in Russian governmental revenues would lead to a heightened state of anxiety that could create a dangerous scenario if the political situation is not handled with great skill. It would actually be ideal if the political situation deteriorated so far that ideological and regional political factions quickly dismembered Russia, but clever intelligence solutions and skillful political overtures would have to be swiftly forthcoming if the nuclear arsenal is to be successfully secured.

However, before such a situation can occur, the Russian political and military leadership are likely to become agitated as they watch the solar power satellites rise in the sky and the value of their vast petrochemical resources in turn largely wiped out. Should the leadership become increasingly menacing, it would be wise to use the prospect of new and large business opportunities in the medium of space to entice the Russians not to take any drastic strategic actions. By doing so, the company can use the opportunity to sow division between Russia and Europe by playing them off against each other in their competition for a carefully calculated pool of space contracts that arose from the deep reservoir of resources that the solar power satellites engendered. And if Russia did disintegrate in a timely fashion, with clandestine operations speeding the process, then the pressure would be off the Company to award the Russians with as large of contracts as would otherwise be granted. Regardless of what happens, with immense economic opportunities opening up in space civilization, the company could entice Russia, or the territories emerging from the former Russia, to re-orient their governance to Western style democracy.

In addition, the joint-stock company could offer to swap a small percentage of its stock for strategic assets in Europe and Russia, perhaps as much as 5 % of the company. As this could amount to $5 trillion, this could be enough for the company to take a controlling stake in the European and Russian economies. These assets would enable the company to dominate the political systems of Europe and Russia. Along with this, the company would commit to a considerable outlay in space on European and Russian soil. This outlay could revitalize the moribund European economy considerably and would significantly increase the dynamism of their society. Obviously the outlay would be tremendous for the Russians as well, allowing them to diversify their economy away from their dependence on petrochemicals.

Asia

The first priority of the company in Asia should be shoring up relations with Japan and South Korea. The sophistication of the technology and business practices of these states is enough to warrant significant business opportunities that the company would be creating. Furthermore, it would be wise to reward such staunch allies of the United States and show the world the benefits of long-term friendship with the United States.

China is an eco-disaster zone. Rivers and cropland across the length and breadth of China are filled with pollutants, while desertification encroaches on cities in the north and west of the country, including Beijing. Smog is without doubt a national nightmare. Coal-fired energy plants continue to litter the industrial landscape, yet factories across China must limit their hours of operation in order to conserve energy. China is clearly desperate for an abundant clean energy solution, as well as a tremendous increase in the sophistication of environmental technology. Fortunately for the Han Chinese, and Americans, the company could be in a position to deliver on such solutions in the form of solar power satellites, as well as the rest of the technological explosion occurring at the company, but not without caveats of course.

There is great sensitivity to embarrassment in China and the century of humiliation (nineteenth century) remains ever present in the minds of the Han Chinese people. China therefore needs to be handled with great historical and cultural delicacy. However, an all-out strategy with the solar power satellites and the company's technological dynamism must be pressed with all due alacrity.

The greatest impact of the solar power satellites may not be political, economic, or even military, but rather cultural, as the Moon has a special place in Chinese culture. "The Chinese Moon Festival is on the 15th of the 8th lunar month. It's also known as the Mid-autumn Festival. Chinese culture is deeply imbedded in traditional festivals. Just like Christmas and Thanksgiving in the West, the Moon Festival is one of the most important traditional events for the Chinese. The Moon Festival is full of legendary stories. Legend says that Chang'e flew to the moon, where she has lived ever since" (About.com 2008). China's lunar exploration vehicles have been named for the Chinese Moon goddess Chang'e. For cultural reasons alone, it is likely that the Chinese will be mesmerized by man-made photovoltaic moons shining brilliantly in the sky as they orbit the Earth. Indeed, the cultural issues are perhaps the most attractive aspects that the solar power satellites have to offer in regard to China. The company can leverage these aspects to great effect in furtherance of its political and economic objectives.

Company officials could use this glorious image of American creative freedom and mathematical excellence to entice the mainland Chinese to desire more information about American society. As with the Russian case, Chinese Communist Party (CCP) autocrats could neither distort the brilliance of the solar power satellites and space civilization as it all shone forth from on high, nor could they ever distort the view that captivated and enamored Chinese had of the heavens. The company would therefore be wise to immediately launch a concerted psy-ops campaign against the CCP as soon as the solar power satellites were in the sky with the express purpose of devastating the Party's authoritarian ideology.

The company's intelligence community should use all avenues available for disseminating information about the solar power satellites and the technological explosion into the Chinese mainland—underground, above ground, internet, even state controlled media, etc. The company could link the economical flow of energy from the solar power satellites to the free flow of information between the United States and the people of the mainland. Should the mainland's population receive less hindered access to information from the United States over the internet and through other media channels, what they should find is a society where big bang innovations spring forth from deep wells of freely expressed creativity which contribute to a flourishing economy and culture that is unhindered by autocratic neurosis. The mainland should find further that the immense comprehensive national power of the United States rests on the very freedoms that Americans would offer to the whole of the world.

The solar power satellites and space civilization would be symbolic of the dynamism of America's liberal capitalist democracy. It all would provide concrete evidence to the entire technology obsessed Chinese population that liberal capitalist democracies are more adept at creating long-term economic growth than are authoritarian systems. This fact would be borne out by the way in which liberal societies are better able to foster the creative energies of their populations by allowing for a wide variety of freedoms and individual liberties, especially freedom of expression. By fostering such freedoms and liberties liberal societies are better able to create cultures of innovation which lead to big bang technological breakthroughs, such as the artificial moons, which ultimately lead to ever enhanced and increased economic growth. For a regime that has staked its entire legitimacy on its ability to generate economic growth, as has the CCP, the dynamism surrounding the situation with the artificial moons and space civilization would bankrupt the CCP's ideology, thus making conditions ripe for a political revolution in China. In this regard, the company's intelligence community, in concert with other American political, business, and scientific leaders, could prove to the population of China that if they had only been a democracy they might have been the ones to have these big bang technological breakthroughs.

If a revolution did become possible, it could either be bottom-up—initiated by the stirrings of the general population—or top-down—initiated by a government in ideological bankruptcy. A revolution initiated by the top would be far more orderly. If the power of the solar power satellites and space civilization were enough that a transitioning leadership sought full alliance with the company while they introduced democratic principles and institutions to Chinese society, it could limit the danger of nationalist hardliners seizing power by military force.

But a revolution initiated by the bottom could result in so much disorder that political control over the whole of the mainland was irrevocably lost and the country could fragment into pieces. It is difficult to say how many pieces the country could fragment into, but chances exist for a range of scenarios: a few large territories, a few large territories and a few small territories, or many small territories. Although, even if the revolution was initiated from the top, the company's intelligence operations could still lead to a situation where the leadership lost control and the country fragmented. What is most important for these efforts, however, is that there should be stirrings throughout the length and breadth of the country and not just in important locations such as Beijing.

As with the Russian case, securing the nuclear arsenal would be of paramount importance. But even if military strong men did seize territories and nuclear weapons, the fragmentation of China would still be ideal for the long-term strategic goals of the United States. While there would be a good possibility that military strongmen could seize enough critically located territory, such as coasts and mountain chains, that they could menace US allies in Asia, in far smaller dominions, these strongmen could still never hope to rival the power of the United States in Asia or on the global level, as is the case with the current ambitions of China.

Should fragmentation occur, company officials should strive to coax the territory of what was formerly China out of a "warring states" paradigm and into a "competing states" paradigm, whereby the smaller nations engage in cutthroat economic competition among each other. The company should obviously press all power aspects of the solar power satellites and the rest of its space activities to great effect in order to limit the strength of strong men that may emerge, and in turn strive to promote democracy among the Han Chinese, doing everything possible to relegate the authoritarian ideology of the former CCP to the dustbin of history.

Whether China remains united as a new democracy, or is disunited in competing democracies, the company should press for significant trade concessions while seeking to open as many markets as possible for all American industries, especially the financial industry. In this regard, the gates to the Chinese economy could be totally flung open to many American businesses. Of most significant importance to the company would be the business opportunities on the Chinese mainland. As the company would be capitalized at up to $100 trillion, it could offer to swap a small percentage of its stock for a considerable amount of China's state owned enterprises. Indeed, a trillion dollars would go a long way in this regard. $5 trillion may be enough for the vast majority of their state owned companies. In this manner, the joint-stock company could gain control of the Chinese economy. This dynamic would allow the company to exert considerable control over the Chinese political system. In exchange for these concessions perhaps the joint-stock company would commit to having a considerable outlay in space on Chinese territory. The techno-explosive effects on China could be striking under such a scenario.

With the fall of the CCP, the regime in North Korea would collapse immediately. If this happens the company should do everything diplomatically possible to ensure that this regime does not lash out in any desperate acts of war, but recognizes that the game is over and proceeds to engage in a peaceful process of reunification with South Korea. The ability of the solar power satellites to shoot down missiles would be extremely persuasive in this regard. The Burmese junta would be under considerable pressure as well following the revolutionary events in China and would likely fall shortly thereafter. Perhaps the Company should send intelligence operatives to facilitate the process and ensure that a new regime arises in Burma that would be very amenable to the company's interests in Asia.

As the regime in India is one of the few democracies in Asia, it would not be toppled by the company. However, much like the Chinese, the Indians would be under heavy pressure in business negotiations to make considerable trade concessions, especially in the financial sector. Should the Indians accede to the company's business demands, they would of course become a minority shareholder in the company and have a significant outlay in space on their territory. The techno-explosive effects of this arrangement would be extraordinary for India as well.

The Middle East

The symbolic power of crystalline Solar Moons would be of considerable aid in the struggle against Islamic extremists emanating from the Middle East who are attempting to wage cosmic religious war against the West. If crystalline moons appeared to pass through the full phases of the Moon every day and every night in the skies above the Middle East, such a display of wonderment could severely compromise the extremist ideologies of terrorist groups and the regimes that covertly and overtly sponsor terrorism. Should the company be responsible for the imagery of glistening crescents across the sky, it would doubtless place heavy pressure on these terror groups and regimes to recalibrate their public messages to account for the beautiful creation that sprung forth from US culture. Certainly these nefarious actors would be hard pressed by their own people to account for the backwardness that they intend for the Islamic world, while the company was simultaneously leading the globe with such sublime progress. Regardless, the population of the Middle East is certain to be spellbound with fear and wonder.

If the company were to hold the region in awe with the solar power satellites, immediately the pro-American regime in Pakistan would be strengthened. Soon the Pakistani military would be exerting themselves far more strenuously in the mountainous border regions between Pakistan and Afghanistan where al Qaeda and the Taliban have long been hiding. It is possible that with all of the socio-economic dynamism and opportunity that the Company's activities would be creating in Pakistan the dynamics of power could shift so heavily that American forces could finally cross the border from Afghanistan and conduct operations to root out the enemy in Pakistan. What is most important for the company's effort, however, is to have a highly skilled psy-ops campaign that could significantly reduce the power of Islamic fundamentalist propaganda and its hold on the people of Pakistan.

The Solar Moon can, indeed, be likened to the shield of Achilles rising to defeat the oil and natural gas reserves of the Middle East, which are the center of gravity for the entire world. The Iranian regime is likely to collapse like a house of cards once the oil and natural gas markets take note and fall precipitously to historic lows. The symbolism of the immensity of the company's power in the sky would surely speed this regime's collapse. In fact, the company's intelligence may not even have to devote very many resources to Iran in order to bring about regime change. It may be enough for the Company's intelligence just to support pro-American forces bubbling up in the Iranian political system in order to accomplish its goals in the country.

In the wake of the new political dynamic spreading in the Middle East, and all of the regime changes that may ensue across the region, the company should follow with a nation building effort. The new political dynamic would create a much more hospitable environment for development. The nation building operation would engage around the Middle East in an effort to develop the woeful economy of the region. In the course of these operations, the company would be investing some of its own resources to stimulate the region's economy. Obviously the company would take strategic stakes in the region's economy that would allow it to dominate the political systems of the region, similar to elsewhere around the world. In addition, as the company will be moving to take control of the global economy with its financial activities on the worldwide equity markets and through all of the stock swaps in conjunction with the global psy-ops campaign, the company will be in position to reorganize global trade and shift some of the production that was previously taking place in industrialized nations onto the territories of the Middle East. The economies of industrialized nations of course will be transformed by the digital galactic complex and their economic production will be increasingly dominated by goods and services for export to the emerging space economy. Exports from the Middle East would surge to industrialized nations while the exports of industrialized nations were surging into space. In this way, the societies of the Middle East could quickly be lifted onto first world industrialized status.

The Global South

At the outset of the global psy-ops campaign, the company should proceed to draw the nations of the global South into its political camp by proffering the construction of space launch facilities in special economic zones across the tropics. The advent of space launch facilities would be a gigantic economic boost to these emerging third world nations and would be a wise investment for a space industry with newly plentiful resources made available by the artificial moons. Each facility could employ tens of thousands of local workers, while the industries feeding services, supplies, and materials into the zone would employ in aggregates of millions. Launch zones and feeder industries located on the equator could be of mammoth size and could potentially employ far more. There could be at least one hundred launch zones across the tropics launching trillions into space on an annual basis.

The tropics are definitely an attractive location for the company's launch facilities due to their proximity to the equator. Due to the Earth's orbital mechanics locations at or near the equator travel faster as the Earth rotates than do locations at more northerly or southerly latitudes. This means that spacecraft launched at or near the equator are already traveling faster toward orbit when they are launched than are spacecraft launched at more northerly or southerly latitudes. Such orbital dynamics necessarily mean that spacecraft launched at or near the equator need less thrust to reach orbit and can therefore carry more cargo. Therefore, these orbital dynamics make the tropics prime locales for the company's prospective launch facilities.

The company could further increase its revenues, soft power, and global clout by first outlining plans for host regimes by which the facility's unsensitive technology can be applied in the local economies—the "spillover"—and then execute these plans. The launch facility could be central to a broader nation building effort that included educational initiatives, natural resource husbandry advisory bodies, financial consulting, security consulting, etc.

There may be some reflexive apprehension about the notion of disseminating launch facility applications, but it should be recognized that unlike with strategic competitors such as China, sharing technology with the nations of the global South would pose little threat to the company's strategic and commercial interests in space or elsewhere. Furthermore, there would be many safeguards if an organizational vehicle was created as part of the zone that would allow company employees to manage the spread of unsensitive technologies in order to ensure they were safely applied in local economies.

Critical applications of the facility and spacecraft would obviously remain secret with very limited access for local personnel for understandable concerns about espionage. But the bottom line is that these facilities would be of tremendous value for a company seeking to launch trillions of dollars into space. This activity would create a far larger market for space goods and services while providing a wide variety of launch options that would break the back of the launch businesses of strategic competitors.

The regimes of host nations may be initially reluctant to forego desires of full or dual control over the special zone, but the prospect of losing the tremendous economic prospects that the zone would offer could be enough to force even the most recalcitrant regimes to acquiesce to the company's zone prerogatives. A skilled global information campaign showcasing the wonder of the artificial moons and advertising the noble ambitions of building space civilization would allow company interlocutors to cloak themselves in righteousness, thereby giving them increased negotiating power should it appear before the local public that the regime's demands were going to break the deal. That the company would be the primary source of funding for the site should provide all the necessary leverage for retaining operational sovereignty within the zone.

Retaining operational sovereignty over the zone is of extreme importance as the zone has tremendous utility as a Trojan Horse if a local regime decides to exhibit bad behavior. First, the facility could be shut down and rendered unusable. Such an action would result in massive job losses at the facility itself and would ultimately destroy the feeder economy, placing heavy pressure on the local regime.

Second, it is reasonable for the company to provide security for the launch facility with methods and means that the company deems appropriate, especially if the company is going to be funding the site. And after 9/11, it is entirely justifiable for these methods and means to include helicopter, drone and fighter patrols, and future combat systems with zone appropriate exteriors. Initially, these zones could be dually used as garrisons to house a small number of politically camouflaged combat forces. Not only would each zone house significant counter espionage operations, but the site could also serve as an entrée for espionage activity throughout the country. Should the local situation deteriorate, additional combat forces could be rapidly deployed to the zone. And if a regime attempted to take over the facility in a bid to nationalize it, instead of rendering the facility unusable, the company could respond by moving to topple the regime. Ideally espionage activity would obtain the support of a split away faction of the enemy military before hostilities commenced. The dynamics for regime change would of course be much more favorable if the regime's actions appeared belligerent before the world. Due to the likelihood of political trouble in some third world launch zones, it would be prudent for the company to have redundant sites scattered throughout the tropics.

Yet even though the zone affords many options for the defense of the company's interests, it need not become a Trojan Horse unnecessarily. Indeed, if handled with the requisite political acumen, the launch facility could very easily enjoy many more diplomatic successes than failures in the global South. In fact it could be the foundation for new security groupings, political forums, and economic organizations between the company and the global South.

By actively promoting economic growth and raising the technological aptitude of emerging and impoverished nations with the launch zone and its technological spillover, the company would go a long way in filling a large bastion of goodwill with the global South. If the company were able to successfully draw the global South into its political and commercial camp then it could steadily tilt the balance of global power by marshaling these regions in support of its international agenda.

Obviously while the company was engaged in building the nations of the global South with the launch facility, it would seek to acquire strategic assets in these regions, as it did elsewhere around the Earth. It could easily do this with the immensity of resources being generated by the space economy as well as energy and technology sales. It would invest these resources in these nations in an effort to build them up onto the level of first world status. Nations in Latin America, especially Brazil, and nations in Southeast Asia would be excellent investment opportunities with significant growth potential. In addition, the company could also offer to swap a small percentage of its stock for strategic assets in these regions that were more established and make these regions minority shareholders in the company. The combination of these investments and asset exchanges would allow the company to dominate the politics of these regions, as it would in a similar manner in the rest of the regions of the Earth. What is more, once the company has taken command of the global economy through its financial activities on the worldwide equity markets in combination with the stock swaps in conjunction with the global psy-ops campaign, the company will be in position to reorganize global trade and shift much of the production from industrialized nations onto the territories of the global South in the same manner as it would in the Middle East. Once again, with the rise of the digital galactic complex the economies of industrialized nations would be transformed to produce exports for the emerging space economy. Such a dynamic would necessarily mean that the global South would then realize significant opportunities to produce more and more goods and services for export to industrialized nations. Thus, the combination of at least one hundred launch facilities across the tropics and the reorganization of global trade shifting much production to the global South would serve to quickly lift the global South onto first world industrialized status.

Chapter 7
Conclusion

The construction of space civilization is a growth prospect. As such, the joint-stock company is something of a growth company, even with a potential valuation of $100 trillion. As space civilization comes into being, and all of its technology spills over across planet Earth, the space and terrestrial assets of the company could eventually become so massive that they would cast a shadow over the entire global economy (quite literally as the artificial moons would eclipse the Sun) (Brown 2012). Indeed, when considering that much of the GDP in a future space civilization would be contained within the company, when considering how sizable will be future energy markets on Earth, when considering how tremendous is the mineral wealth in the cosmos, when considering how great could be future technology sales on Earth, and when considering how significant would be the company's equity and real estate holdings on Earth, it is entirely possible that the space joint-stock company's market value could surge over a quadrillion dollars in the decades following its establishment.

In any event, the first division would be totally prepared to protect the company's interests and possessions, most definitely in space but also on Earth. Ideally it would have a passive role beyond securing the company's constellation of energy in orbit while the activities of the second, third, and fourth divisions flourished. Indeed, all aspects of the DIME (diplomatic, information, military, economic) would be brought to bear around the world to create the most conducive political conditions for the company's vitality. In one way or another—with the fourth division's financial strategy on the global stock exchanges, or the psy-ops campaign leading to stock swaps, or both—the company would arrive at a situation in which it was in complete command of the global economy. This dynamic would allow it to dominate all the political systems of the Earth. As such, the company could be a vehicle for creating the first empire to not only span the whole of the globe, but also to include multiple planets. A multi-planet empire of this wealth and power would be greater even than Rome.

T. Brown, *The Digital Galactic Complex*, SpringerBriefs in Political Science 30,
DOI 10.1007/978-3-319-22386-5_7

The core mission of the joint-stock company would be to build space civilization. This could be the greatest undertaking in the history of the world. Not only would it initiate an explosion in science and medicine, but also it would create tremendous prosperity and opportunity for the whole of the globe. The Solar Moons would be the greatest feat of science and engineering in history and would have a profound impact on humanity. Whatever happens, the joint-stock company would be a fantastic opportunity for Silicon Valley, the American aerospace industry, and NASA. The riches of space civilization are beckoning.

References

Obama administration. National Space Policy. 2010.
Smith, Marcia S. "President Obama's National Space Policy: A change in tone and a focus on space sustainability." *Space Policy* 27 (2011): 20–23.
Kueter, Jeff. "Evaluating the Obama National Space Policy: Continuity and New Priorities." *George C. Marshall Institute*, July 2010: 1–29.
Bush administration. National Space Policy. 2006.
Sadeh, Eligar. "Space policy challenges facing the Barack Obama administration." *Space Policy* 25 (2009): 109–116.
Lichodziejewski, David, Gordon Veal, Richard Helms, Robert Freeland, and Mark Kruer. "INFLATABLE RIGIDIZABLE SOLAR ARRAY FOR SMALL SATELLITES." *American Institute of Aeronautics and Astronautics*, 2002. http://www.dtic.mil/dtic/tr/fulltext/u2/a451501.pdf. Accessed 18 April 2013.
Garretson, Peter. "Solar Power in Space?" *Strategic Studies Quarterly*, Vol. 6 No. 1 (Spring 2012) 123.
Schauer, William H. *The Politics of Space*. New York: Holmes and Meier Publishers, Inc., 1976.
Clausewitz, Carl von. *On War*, Book 6. Translated by Michael Howard and Peter Paret. New York: Princeton University Press, 1976.
Coletta, Damon, and Frances T. Pilch, editors. *Space and Defense Policy*. London: Routledge, 2009.
Gray, Colin. *Modern Strategy*. Oxford: Oxford University Press, 1999.
Johnson, Nicholas. *Soviet Military Strategy In Space*. London: Jane's Publishing Company Limited, 1987.
Sheldon, John B. "Selling US Space Power Short," *The Space Review*, 4 September 2007. http://www.thespacereview.com/article/948/1. Accessed 9 September 2013.
Hildreth, Steven A., Allison Arnold. "Threats to U.S. National Security Interests in Space: Orbital Debris Mitigation and Removal." Congressional Research Service 8 January 2014. http://fas.org/sgp/crs/natsec/R43353.pdf. Accessed 25 September 2014.
Gruss, Mike. "Lockheed Martin Lands $914M Space Fence Contract." *Space News* 2 June 2014. http://spacenews.com/40776lockheed-martin-lands-914m-space-fence-contract/. Accessed 28 February 2015.
National Aeronautics and Space Administration. "Flurry of Small Breakups in First Half of 2014." Orbital Debris Quarterly News Volume 18, Issue 3. http://orbitaldebris.jsc.nasa.gov/newsletter/pdfs/ODQNv18i3.pdf. Accessed 25 September 2014.
Nelson, Timothy G. "Regulating the Void: In-Orbit Collisions and Space Debris." The Space Review, 27 May 2014 http://www.thespacereview.com/article/2520/1. Accessed 21 July 2014.

T. Brown, *The Digital Galactic Complex*, SpringerBriefs in Political Science 30,
DOI 10.1007/978-3-319-22386-5

Brown, Trevor. "Space and the Sea: Strategic Considerations for the Commons." Astropolitics Vol 10 No 3 (November 2012): 234–247.

Anzaldua, Al, David Dunlop, and Brad Blair. "Are Solar Power Satellites Sitting Ducks for Orbital Debris?" The Space Review 22 September 2014. http://www.thespacereview.com/article/2602/1. Accessed 5 October 2014.

Jennings, Katie. "What if Giant Space-Based Solar Panels Could Beam Unlimited Power to the Earth?" Yahoo Finance 11 August 2014. http://finance.yahoo.com/news/giant-space-based-solar-panels-174349042.html. Accessed 1 September 2014.

Space Technology Mission Directorate. "SPS Alpha: The First Practical Solar Power Satellite Via Arbitrarily Large Phased Array." NASA http://www.nasa.gov/directorates/spacetech/niac/mankins_sps_alpha.html. Accessed 4 October 2014.

Brown, Trevor. "SSP: A Spherical Architecture." The Space Review 1 June 2009. http://www.thespacereview.com/article/1383/1. Accessed 10 May 2013.

McDuffee, Allen. "The Navy's Plan to Beam Down Energy from Orbiting Solar Panels." Wired 17 March 2014. http://www.wired.com/2014/03/space-solar/. Accessed 20 September 2014.

"Inflatable Solar Array Technology Packs Incredible Power in Small Package." NASA. http://www.nasa.gov/centers/marshall/news/news/releases/2014/14-112.html. Accessed 1 September 2014.

Brown, Trevor. "A Space Joint Stock Company." The Space Review 30 April 2012. http://www.thespacereview.com/article/2073/1. Accessed 10 May 2013.

Programs: Space Structures. L'Garde. http://www.lgarde.com/space-structures.php. Accessed 1 September 2014.

Brown, Trevor. "An Emerging Concept for the Medium of Space." Comparative Strategy Vol 30 No 4 (September 2011): 373–380.

Brown, Trevor. "Space: The Environmental Frontier." National Strategy Forum Review, (November 2010).

"Solar Stocks: Does the Punishment Fit the Crime?" 24/7 Wall St. 6 October 2011. http://247wallst.com/energy-business/2011/10/06/solar-stocks-does-the-punishment-fit-the-crime-fslr-spwra-stp-jaso-tsl-ldk-tan/. Accessed 7 October 2014.

"The World's Billionaires." Forbes. http://www.forbes.com/billionaires/. Accessed 30 November 2014.

"The Moon Festival," About.com: Chinese Culture, http://chineseculture.about.com/library/weekly/aa093097.htm. Accessed 12 December 2008.

Dreier, Casey. "A Generational Opportunity for Europa." *The Space Review*, 21 July 2014 http://www.thespacereview.com/article/2561/1. Accessed 21 July 2014.

"Feasibility Study of Commercial Space Manufacturing, Phase II Final Report," Volume I: Executive Summary, MDC E1625, McDonnell Douglas Astronautics Company, East St. Louis, Missouri, January 15, 1977, pp. 1–2, 8–20.

"Space Industrialization: Final Report," Volume 1. Executive Summary, SD 78-AP-0055-1, Rockwell International Space Division, Contract NAS8-32198, April 14, 1978, pp. 1–8.

"Feasibility of Commercial Space Manufacturing: Production of Pharmaceuticals," Final Report, Volume 1, Executive Summary, MDC E2104, McDonnell Douglas Astronautics Company, St. Louis Division, Contract NAS8-31353, November 9, 1978, pp. 1–3, 26–30.

James Beggs, Administrator, NASA, to William Clark, Assistant to the President for National Security Affairs, August 26, 1983, with attached: John F. Yardley, President, McDonnell Douglas Astronautics Company, to James Beggs, Administrator, NASA, August 23, 1983.

"The Printed World." *The Economist* 10 February 2011. http://www.economist.com/node/18114221. Accessed 24 August 2014.

Hart, Brad. "Will 3D Printing Change the World?" *Forbes* 6 March 2012. http://www.forbes.com/sites/gcaptain/2012/03/06/will-3d-printing-change-the-world/. Accessed 24 August 2014.

Tate, Karl. "Home on the Moon: How to Build a Lunar Colony." *Space.com* 25 June 2013.
http://www.space.com/21588-how-moon-base-lunar-colony-works-infographic.html. Accessed 24 August 2014.
Moskowitz, Clara. "Shadows of the Moon Hide Fluffy Dirt and Water Ice." *Space.com* 19 January 2012. http://www.space.com/14284-moon-permanently-shadowed-regions-water-ice.html. Accessed 24 August 2014.
Dinkin, Sam. "Colonize the Moon Before Mars." *The Space Review* 7 September 2004.
http://www.thespacereview.com/article/221/1. Accessed 24 August 2014.
Garretson, Peter. "Asteroid Strike! Asteroid Mining! Will the Air Force Have a Role?" Air and Space Power Journal 2013 http://www.au.af.mil/au/afri/aspj/article.asp?id=174. Accessed 4 October 2014.
"Near Earth Object Program." Jet Propulsion Laboratory. http://neo.jpl.nasa.gov/neo/. Accessed 26 August 2014.
"Near Earth Object Program." Jet Propulsion Laboratory. http://neo.jpl.nasa.gov/neo/resource.html. Accessed 26 August 2014.
"Here's Why Asteroid Mining Will Fuel Human Expansion into the Cosmos." Planetary Resources. http://www.planetaryresources.com/2013/11/heres-why-asteroid-mining-will-fuel-human-expansion-into-the-cosmos/. Accessed 26 August 2014.
Larson, K. Dean. "Mining Asteroids and Exploiting the New Space Economy." *The Wall Street Journal* 21 August 2014. http://online.wsj.com/articles/dean-larson-mining-asteroids-and-exploiting-the-new-space-economy-1408662987. Accessed 26 August 2014.
Tate, Karl. "How Asteroid Mining Could Work." *Space.com* 22 January 2013. http://www.space.com/15391-asteroid-mining-space-planetary-resources-infographic.html. Accessed 26 August 2014.
Thomas, Susan. "Gold Rush in Space? Asteroid Miners Prepare, but Eye Water First." *Reuters* 21 November 2013. http://www.reuters.com/article/2013/11/21/us-space-mining-asteroids-idUSBRE9AK0JF20131121. Accessed 26 August 2014.
Godwin, Richard. "New Fort Knox: A Means to a Solar-System-Wide Economy." *The Space Review*, 21 July 2014 http://www.thespacereview.com/article/2558/1. Accessed 21 July 2014.
Mack, Erick "Elon Musk on the Details of Colonizing Mars." *Forbes* 12 December 2013. http://www.forbes.com/sites/ericmack/2013/12/12/elon-musk-on-the-details-of-colonizing-mars/. Accessed 26 August 2014.
Vergano, Dan. "Will We Be the First Martians?" *USA Today* 13 September 2012. http://usatoday30.usatoday.com/tech/science/story/2012-09-04/Mars-curiosity-future-life/57778412/1. Accessed 26 August 2014.
Moskvitch, Katia. "Life on Mars: How a One-Way Martian Colony Project Could Work." *Space.com* 4 December 2013. http://www.space.com/23838-mars-one-colony-martian-volunteers.html. Accessed 26 August 2014.
Hertzfeld, Henry R. "Space as an Investment in Economic Growth." P. 385-401. Chapter Three in *Exploring the Unknown: Selected Documents in the History of the U.S. Civil Space Program.* Volume III: Using Space. Edited by John M. Logsdon with Roger Launius, David H. Onkst, and Stephen J. Garber. Washington, D.C.: National Aeronautics and Space Administration, NASA History Division, Office of Policy and Plans, 1998.
Roger W. Hough, "Some Major Impacts of the National Space Program," Stanford Research Institute, Contract NASW-1722, June, 1968, pp. 1–2, 19–22, 36.
"Economic Impact of Stimulated Technological Activity," Final Report, Midwest Research Institute, Contract NASW-2030, October 15, 1971, pp. 1–11.
Michael K. Evans, "The Economic Impact of NASA R&D Spending," Executive Summary, Chase Econometric Associates, Inc., Bala Cynwyd, Pennsylvania, Contract NASW-2741, April 1976, pp. i–iii, 1–18.
"Economic Impact and Technological Progress of NASA Research and Development Expenditures," Executive Summary, Midwest Research Institute, Kansas City, Missouri, for the National Academy of Public Administration, September 20, 1988, pp. 1–4.

Martin D. Robbins, John A. Kelley, and Linda Elliot, "Mission-Oriented R&D and the Advancement of Technology: The Impact of NASA Contributions," Final Report, Industrial Economics Division, Denver Research Institute, University of Denver, Contract NSR 06-004-063, May 1972, pp. iii–iv, 25–39, 59.

"Quantifying the Benefits to the National Economy from Secondary Applications of NASA Technology—Executive Summary," NASA CR-2674, Mathematica, Inc., March 1976.

"Economic Effects of a Space Station: Preliminary Results," NASA, June 16, 1983, pp. 1–2, 20–21.

"The Economic Impact of the Space Program: A Macro and Industrial Perspective," prepared for Rockwell International by The WEFA Group, Bala Cynwyd, Pennsylvania, May 1994, pp. 1–4.

Craig L. Fuller, The White House, Memorandum for the Cabinet Council on Commerce and Trade, "Commercial Space Initiatives," April 10, 1984, with attached: "Private Enterprise in Space—An Industry View," pp. iv–v.

"Space Industrialization: Final Report," Volume 1. Executive Summary, SD 78-AP-0055-1, Rockwell International Space Division, Contract NAS8-32198, April 14, 1978, pp. 1–8.

"Space Industrialization: An Overview," Final Report, Volume 1, SAI-79-662-HU, Science Applications, Inc., April 15, 1978, pp. 1–5, 10–12, 15–17.

Leo S. Packer, Special Assistant to Associate Administrator, Office of Advanced Research and Technology, NASA, "Proposal for Enhancing NASA Technology Transfer to Civil Systems," September 26, 1969, pp. 1–9.

Richard L. Chapman, Loretta C. Lohman, and Marilyn J. Chapman, "An Exploration of Benefits from NASA 'Spinoff,'" Chapman Research Group, Contract 88-01 with NERAC, Inc., June 1989, pp. 1–5, 23–28.

H. R. Hertzfeld, "Technology Transfer White Paper," internal NASA document, June 23, 1978.

Leffler, George L. The Stock Market. New York: The Ronald Press Company, 1963.

Bolten, Steven E. Stock Market Cycles: A Practical Explanation. Westport, Connecticut: Quorum Books, 2000.

Marathe, Achla. Studies on the Behavior of Equity Markets. New York: Garland Publishing, Inc., 1998.

Dunn, Ruben J. and John Morris. The Crash Put Simply. New York: Praeger, 1988.

Board of Governors. The Federal Reserve System: Purposes and Functions. Washington, D.C.: Federal Reserve, 1963.

FSC
www.fsc.org
MIX
Papier | Fördert
gute Waldnutzung
FSC® C083411

Kolibri 360 GmbH
Ferdinand-Jühlke-Straße 7
99095 Erfurt, Deutschland
produktsicherheit@kolibri360.de